Bill B. Dean, PhD

Managing the Potato Production System

*Pre-publication
REVIEWS,
COMMENTARIES,
EVALUATIONS . . .*

"THIS IS THE BEST COMPREHENSIVE TEXT AVAILABLE ON POTATO CROP MANAGEMENT, and this book will be an invaluable aid for students, producers, crop consultants, and especially for those laboratory scientists who don't know beans about spuds. This is not an exhaustive scientific review, but rather a comprehensive guide to all facets of potato production from seed handling to field management, pest control, harvest, and storage. The text is easy to read, has good illustrations, and is sprinkled with interesting facts and statistics. Additional information is provided on the history of the potato, explanation of breeding systems, cultivar characteristics, etc.

I have been teaching a potato crop management course at the University of Idaho for the past eight years, and this is the first reference I've seen that can serve as a complete textbook for this class. I intend to use it."

Robert B. Dwelle, PhD
*Professor of Crop Physiology,
University of Idaho*

Food Products Press
An Imprint of The Haworth Press, Inc.

Managing the Potato
Production System

FOOD PRODUCTS PRESS

An Imprint of The Haworth Press, Inc.
Robert E. Gough, PhD, Senior Editor

New, Recent, and Forthcoming Titles:

The Highbush Blueberry and Its Management by Robert E. Gough

Glossary of Vital Terms for the Home Gardener
by Robert E. Gough

Seed Quality: Basic Mechanisms and Agricultural Implications
edited by Amarjit S. Basra

Statistical Methods for Food and Agriculture edited by Filmore
E. Bender, Larry W. Douglass, and Amihud Kramer

World Food and You by Nan Unklesbay

Introduction to the General Principles of Aquaculture
by Hans Ackefors, Jay V. Huner, and Mark Konikoff

Managing the Potato Production System by Bill B. Dean

Marketing Livestock and Meat by William Lesser

The World Apple Market by A. Desmond O'Rourke

*Understanding the Japanese Food and Agrimarket: A Multifaceted
Opportunity* edited by A. Desmond O'Rourke

Marketing Beef in Japan by William A. Kerr et al.

Managing the Potato Production System

Bill B. Dean, PhD

Food Products Press
An Imprint of The Haworth Press, Inc.
New York • London • Norwood (Australia)

Published by

Food Products Press, an imprint of The Haworth Press, Inc., 10 Alice Street, Binghamton, NY 13904-1580

Library of Congress Cataloging-in-Publication Data

Dean, Bill.
    Managing the potato production system / Bill Dean.
        p.   cm.
    Includes bibliographical references and index.
    ISBN 1-56022-025-2 (acid free paper).
    1. Potatoes. I.Title

SB211.P8D32 1993
635'.21–dc20   92-1676
                                                                           CIP

CONTENTS

ABOUT THE AUTHOR

Bill B. Dean, PhD, is Associate Horticulturist and Extension Specialist at Washington State University in Richland. As Director of Agronomy for Agri Northwest (UI Group), Dr. Dean developed production management systems with several staff experts for the purpose of managing 14,000 acres of potatoes produced in the Columbia Basin of Washington. He has done research on vegetable crops, the effects of environment on potato production, and the marketing of vegetable crops for export. He is a member of the American Society for Horticultural Science, the Potato Association of America, and the American Society of Plant Physiologists.

List of Figures

List of Tables

Foreword

Near the end of the sixteenth century, European explorers returning home carried strange foods with them from the New World. Among those foods were the Jerusalem artichoke and the potato. Both were widely accepted by native peoples, and the potato already had been cultivated in its high Andean homeland for 40 centuries. While the Jerusalem artichoke was readily accepted by the Europeans as food "fit for the queen," they heaped scorn upon the potato. Being a member of the deadly nightshade family, it was thought to be poisonous. A minister of the church even preached against it, reasoning that if God had meant for man to eat the potato, the plant would have been mentioned in the Bible. War-ravaged Europe slowly came to understand the blessings of this plant. While grains were burned and trampled beneath the boots of advancing armies, the potato's tubers, safely underground, remained undamaged. It grew nearly anywhere and could be stored for long periods of time. Particularly in Ireland, it became the crop of the peasantry. Indeed, the Irish poor lived almost exclusively on a diet of potatoes, consuming 10-12 pounds per person each day from the beginning of the seventeenth century until the great famine struck in 1845. That amount more than satisfied a person's daily requirement of calories, protein, iron, and vitamins B and C, and provided half the phosphorous and a tenth the calcium. Adding milk to the potato rounded out the dietary requirements and so preserved the Irish people. Similary, the potato kept the German people alive through two world wars and is recognized today as the valuable food that it is.

This book is about the culture of the potato. It contains the latest scientific information in a language and form the grower can easily understand. For unless the grower can understand the findings of the researcher, the grower cannot implement them and they remain only academic doodlings in obscure journals. Americans help feed the world and have the greatest system of agriculture in the world today.

They arrived at this position of strength through diligence in making fundamental advances in science and technology that were readily accepted and implemented by progressive growers. American farming has advanced constantly and was not, nor is not presently interested in maintaining the status quo, for it is only by progressing that farming remains alive. A plant that grows remains productive and fruitful. When it stops growing, it weakens and eventually dies.

Some of the old methods are still valuable and valid in today's agriculture, but many have been superceded. New potatoes and new ways to grow and market them have been developed. This remarkable book, written for the most important person in potato production—the grower—contains all the current information needed to harvest a bountiful crop at the dawn of the twenty-first century.

R. E. Gough
Senior Editor
Food Products Press

Preface

This book is written for potato producers and those who help them. It is not meant to be an exhaustive dissertation about potatoes. There have been several excellent books written recently that convey what scientists know about potatoes. Rather than restate what others have said, I have attempted to impart my understanding of what is scientifically known, toward the goal of managing the potato production system under irrigation.

There are many areas where absolute knowledge about potato production is lacking and therefore definitive advice cannot be provided. There are other areas where experimental evidence is weak but conclusions may be made to reduce risks within the system. In addition, intuition has been used in places where application of knowledge from distant sources may provide reasonable assurance that a decision is correct. Of course, there are volumes of information that expound basic scientific truths for us to utilize. While there are always skeptics, even of hard facts, we must be fair in our attempts to discern the truth by rigorously testing experimental results in real situations. We may find that although some truths are unquestionable, they have little impact on the production system, while others may be profound in their effects.

Successful producers are individuals who can search out the truths of potato growing, achieve the application of major truths rapidly, and incorporate the finer points over time. The truisms may be obtained from scientists using experimental approaches, from producers using sound reasoning and experiences, or from consultation with others who have honestly sought the truth. Validation of true production practices ultimately rests in the hands of producers who test procedures over many years and under variable conditions. Producers must continually evaluate the procedures they use in light of today's knowledge and with as much foresight as they are given. Today, we do not see clearly where we are headed, but we can gain glimpses if we watch and keep our minds open. When we arrive at a point in the future that we shared a glimpse of in the past, we will begin to understand how much we can trust our intuition and ability to see what is ahead.

We cannot be afraid to change. Many of the practices presented in this book have already been tested extensively and proven to be valuable. This does not mean that there are no situations where they will fail or that they are sound in all areas of economic, environmental, and physical practices. Producers must evaluate each practice in light of their own situations. There is no single production system that will function well with the widely diverse production regions, geography, and climate types found in the United States. My attempt has been to teach principles of potato production via existing knowledge of how the crop grows and what influences its production as a crop. Producers, advisors, or consultants must evaluate their own situations and then test the procedures in those situations. To this end I hope that the information is clearly presented and properly documented so that it will be a useful resource.

I would like to thank those people who have taught me much about potatoes during the past 20 years. They include my major professors, Doyle Smittle, Robert Kunkel, and Pappachan Kolattukudy, as well as good friends and colleagues, Robert E. Thornton and Dennis Corsini. I am also grateful to those who enhanced my understanding of the practical aspects of growing potato seed, such as John Scutter, Sr. and William Kimm, as well as others in the Montana Potato Seed Association. Special thanks are due to those who trusted and encouraged me at Agri Northwest (UI Group) during the period of implementing the potato production system on their farms, specifically Martin Wistisen, Kent Nielsen, Larry Hector, Jeff Mason, and Rodney Larson, as well as their managerial staffs. I cannot separate my learning from those who have worked with or for me because they provided much of the testing of our hypothesis. Key people who have provided technical support are Norris Holstad and Robert K. Thornton, without whom I would not have achieved much. I would also like to acknowledge the efforts of Robert E. Thornton in his review and suggestions for this book. Special thanks are due to the Washington State University Cooperative Extension artists for the drawings for Figures 5, 6, 7, 10, 12, and 16, and to Matthew W. Dean for Figures 3, 4, 8, and 9.

With these few opening remarks, I would like to say that as in any venture, success will be available to those who can assimilate information and apply it without significant mistakes. We shall all find situations in which our information is incomplete and we must rely on others. When we learn to accept other's ideas and work together, a system develops. The outcome of working together and sharing each others' expertise toward a common goal is a far more rewarding experience. My goal in writing this text is to help potato producers grow an economically viable crop to provide food for consumers and to do it in a manner that can be sustained over generations with positive impacts on the environment. To that end, I hope we can all work together with a competitive but synergistic spirit.

Chapter 1

History and Marketing

INTRODUCTION

The scientific (Latin) name for the common cultivated potato has three parts: the family, Solanaceae; the genus, *Solanum;* and the species, *tuberosum.* Eight species of the cultivated potato of the genus *Solanum* exist (Table 1). Though they may not seem important to producers, they are an essential part of potato breeding programs throughout the world. They provide genetic characteristics that can be incorporated into new cultivars (cvs, cultivated varieties). In addition, there are at least 154 other species of *Solanum* which are found in the areas where potatoes originated. As scientists unravel the genetics of these species and are able through conventional or new breeding techniques to incorporate their desirable characteristics (genes), fundamental changes in potato production will occur. As an example, when resistance to late blight was incorporated into potatoes in Europe, famines like those that occurred in the 1700s and 1800s were no longer a major concern. Significant advances in genetic improvement in potatoes are anticipated within a short period, particularly with regard to virus, insect, and nematode resistances. These improvements will occur as the basic genetic information obtained from studying the pests and the natural resistance found in the *Solanum* gene pool (wild and cultivated potatoes) is utilized.

The cultivated potato had its origin in the Andean mountains of South America. Some other *Solanum* species have been shown to have originated in areas further north through Central America and into North America (Correll, 1962; Medsger, 1939). The International Potato Center (CIP) was established in Lima, Peru, with one of its purposes being to gather and maintain wild and cultivated potatoes from these areas for use in developing new potato cultivars. This center provides basic genetic material and training for people around the world, and is responsible for preventing the loss of this valuable genetic resource. The breeding programs in the United States obtain desirable species from this resource base through the plant introduction station in Sturgeon Bay, Wisconsin.

History and Marketing

The history of the potato in North America is somewhat clouded. It was probably introduced to colonists from England via Bermuda in 1621 (Hawkes, 1978). Although potatoes as a crop originated and were grown in South America, they had to travel to Europe, be accepted there as a food crop, and then return to North America for introduction.

The location of potato production in the United States has changed based on consumer needs, new markets, and technological developments. During the period when fresh potato consumption was the primary use for this crop, and transportation was slow or unavailable, the production regions were located in the vicinity of large towns and developing cities. Following the development of rail and truck transportation and processing techniques which required particular raw product characteristics, production in more distant, climatologically favorable regions began to expand (Table 2). The economical aspects of production, processing, and marketing have also played an important role in development of potato production regions.

A significant per unit production cost advantage exists in the Northwest because of high per acre yields and relatively low per unit input costs (Greig and Blakeslee, 1988). The results of the ability of producers in different regions to utilize climatic advantages and new technologies is illustrated by a comparison of per acre yields for three different production regions compared to the average U.S. production (Figure 1). In Washington State, potatoes have been produced primarily in the climatically favorable central Columbia Basin area since the completion of an irrigation project

TABLE 1. Two classification schemes of cultivated potato (*Solanum*) species.

Hawkes[x]	Correll[y]	Sets of chromosomes (ploidy)	Characteristics and adaptation
S. ajanhuiri	Group *stenotomum*	2n = 24	Small blue flower, frost resistant, S. Peru and N. Bolivia, high altitude
S. goniocalyx		2n = 24	White or pink flower, tubers bright yellow flesh, C. to N. Peru, high altitude
S. phureja	Group *phureja* Subgroup *amarilla goniocalix*	2n = 24	Tubers in 3-4 months, no dormancy, adapted to drought and frost-free areas, wet mountain slopes of eastern Andes
S. stenotomum		2n = 24	Tubers in 5-6 months, long dormancy, some forms frost resistant, C. Peru to N. Boliva
S. x chaucha	Group *chaucha*	2n = 36	C. Peru to C. Bolivia, high altitudes
S. x juzepczukii	*S. X Juzepczukii*	2n = 36	Semi-rosette habit, frost resistant C. Peru to S. Bolivia, high altitude
S. tuberosum Subspecies *tuberosum*		2n = 48	Tubers- long dormancy, S. Chile, worldwide
Subspecies *andigena*	Group *andigena*	2n = 48	Ancestral subspecies of *S. tuberosum*
S. x curtilobum	Group *tuberosum*	2N = 60	Semi-rosette habit, frost resistant, variation in tuber color, C. Peru to S. Bolivia, very high altitude

[x]Hawkes, J. G. 1978 "History of the Potato." In *The Potato Crop*, edited by P. M. Harris. Chapman and Hall, London, p. 13.
[y]Correll, D. S. 1962. *The Potato and Its Wild Relatives*. Texas Research Foundation, Renner, Texas. 606 pp.

TABLE 2. Historical potato production in the United States (1,000 acres). (From USDA *Agricultural Statistics*, yearly volumes)

State	1928-32	1935	1940	1945	1950	1955	1960	1965	1970	1975	1980	1985	1989
Alabama	31	38	51	47	32	26	22	20	17	20	15	13	13
Arizona	3	2	2	6	5	5	10	11	11	6	4	6	6
Arkansas	35	44	41	38	20	11	6	3	1	–	–	–	–
California	38	45	72	119	122	115	104	107	90	61	57	62	49
Colorado	99	84	67	91	56	52	56	48	37	40	43	63	68
Connecticut	12	15	16	20	9	7	7	7	5	2	2	1	<1
Florida	28	26	30	35	25	38	37	41	36	28	28	35	43
Georgia	15	21	24	22	9	6	2	1	–	–	–	–	–
Idaho	99	89	128	200	164	170	234	282	327	312	300	345	353
Illinois	50	50	39	19	8	4	–	2	2	2	2	3	3
Indiana	55	66	51	29	17	10	7	8	7	7	5	5	4
Iowa	74	84	60	25	11	6	4	3	3	3	1	2	2

Year

State	1928-32	1935	1940	1945	1950	1955	1960	1965	1970	1975	1980	1985	1989
Kansas	45	35	25	17	8	3	2	1	1	–	–	–	–
Kentucky	53	67	44	38	22	17	11	5	3	–	–	–	–
Louisiana	38	42	40	42	14	10	4	4	3	3	2	1	<1
Maine	175	161	157	209	132	141	147	148	150	122	104	99	80
Maryland	32	33	20	18	10	4	4	3	3	2	2	2	2
Massachusetts	12	16	18	22	11	7	7	7	5	4	3	3	3
Michigan	247	263	214	164	85	58	47	48	40	37	40	58	40
Minnesota	353	334	250	174	93	81	105	89	97	65	64	76	69
Mississippi	11	16	22	26	12	10	4	3	3	2	–	–	
Missouri	54	54	41	26	16	9	5	4	1	–	–	–	
Montana	21	23	16	17	12	9	8	8	8	8	7	7	8
Nebraska	117	126	81	69	43	20	11	11	7	8	8	11	10
Nevada	3	3	2	4	2	2	1	1	–	13	13	9	8
New Hampshire	9	10	8	7	4	3	2	1	1	<1	–	–	–

TABLE 2 (continued)

State	Year												
	1928-32	1935	1940	1945	1950	1955	1960	1965	1970	1975	1980	1985	1989
New Jersey	42	50	55	71	38	22	19	17	12	7	8	9	5
New Mexico	5	7	3	5	1	1	2	2	3	4	3	10	12
New York	211	195	199	182	110	97	87	75	65	47	44	38	29
N. Carolina	74	83	80	72	62	37	29	16	15	16	17	14	17
N. Dakota	125	135	162	170	112	87	112	106	116	110	112	139	137
Ohio	111	121	92	59	30	23	16	16	15	12	11	9	8
Oklahoma	43	40	30	18	9	5	2	1	–	–	–	–	–
Oregon	38	37	35	52	38	25	47	39	55	56	47	61	50
Pennsylvania	196	198	168	137	83	58	36	38	35	29	22	22	21
Rhode Island	2	4	5	7	4	5	5	6	5	4	3	3	1
S. Carolina	22	18	25	21	15	9	7	2	–	–	–	–	–
S. Dakota	60	50	30	31	14	10	7	5	8	5	7	12	9
Tennessee	46	57	44	35	23	15	9	6	4	5	3	3	<1
Texas	54	54	52	49	27	18	19	21	26	15	14	20	16

| | Year | | | | | | | | | | | | | |
State	1928-32	1935	1940	1945	1950	1955	1960	1965	1970	1975	1980	1985	1989
Utah	13	14	13	18	14	9	9	9	6	6	5	7	6
Vermont	16	16	13	10	5	3	2	2	1	1	1	<1	–
Virginia	111	88	74	68	44	36	29	27	32	25	14	17	12
Washington	46	38	39	38	31	38	35	52	87	105	87	126	118
West Virginia	39	34	33	29	17	13	10	5	5	4	–	–	–
Wisconsin	251	253	179	128	67	52	55	59	37	50	50	64	68
Wyoming	26	27	12	14	8	6	4	4	4	7	6	1	2
Total Acres	3,243	3,271	2,886	2,729	1,713	1,452	1,410	1,419	1,449	1,299	1,175	1,409	1,272

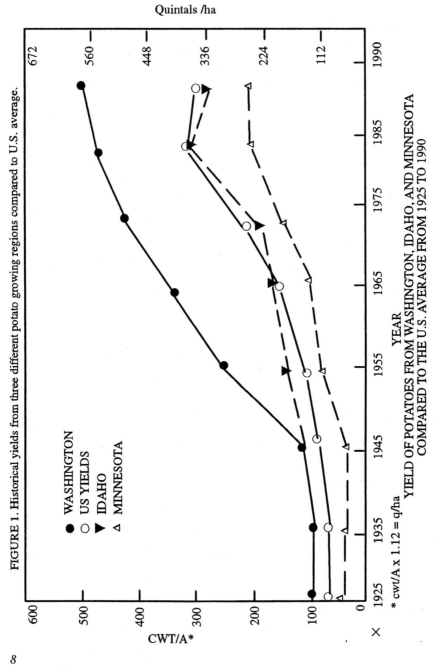

FIGURE 1. Historical yields from three different potato growing regions compared to U.S. average.

Quintals /ha

YEAR

YIELD OF POTATOES FROM WASHINGTON, IDAHO, AND MINNESOTA
COMPARED TO THE U.S. AVERAGE FROM 1925 TO 1990

* cwt/A x 1.12 = q/ha

WASHINGTON
US YIELDS
IDAHO
MINNESOTA

CWT/A*

8

by the Federal Bureau of Reclamation in the late 1940s. Subsequent improvements in cultural practices are responsible for the substantial gains in per acre yield between 1950 and 1980. In Minnesota, the majority of potatoes are produced without irrigation, which is the main yield constraint. Much of Idaho's production is from high mountain valleys or desert plateaus where the length of the growing season limits yield.

The production of potatoes in the sparsely populated West became more dominant as processing techniques were developed during the mid-1900s. Innovative marketing strategies as well as new consumer demands for processed products also influenced the rapid expansion of potato production in the West.

The following is an account of the development of the Northwest frozen potato products industry as reported in *The Diamond Book* (1989).

Probably the most significant food processing developments of the 1950s were centered around the explosive growth of the potato as a frozen food item. What began as the decade of the frozen pea ended as the decade of the french fry. Responding to insatiable demand, the Northwest frozen potato pack grew at an unparalleled rate–from 6 million pounds in 1953 to 262 million pounds in 1959. With about 75 percent of the nation's production, the Northwest was clearly established as the dominant frozen potato producing region.

Idaho's J. R. "Jack" Simplot is considered one of the pioneers in the evolution of frozen french fries. Simplot began freezing potatoes at his Nampa, Idaho, plant in the late 1940s. Based on research by Ray Dunlap, the J. R. Simplot Company was the first processor to produce an oil-blanched, par-fried product that was particularly adaptable to the fast food and institutional food services markets.

Directed by F. Nephi Grigg, Ore-Ida Foods began its rapid development in 1953 at Ontario, Oregon. Ore-Ida introduced potato patties and the now-famous "tater tot." The company is also credited with developing a new method of freezing potatoes.

In 1960, Lamb-Weston, Inc., joined Simplot and Ore-Ida in a triad destined for world leadership in the frozen potato business. It was then that F. Gilbert "Gib" Lamb and his chief engineer, Art Davidson, invented the Lamb Water Gun Knife. This innovation introduced a highly efficient concept that is still employed today. A high-velocity stream of water propels whole potatoes through a grid of cutting knives oriented to produce uniform potato strips of maximum length and, therefore, maximum value.

Gib Lamb, along with brothers Reese and Paul, either built or bought plants in all three states in the 1960s. With the purchase of Snowflake Canning Company of Brunswick, Maine in 1965, Lamb-Weston added the expertise of two other frozen french fry pioneers, John L. Baxter Jr. and Francis Saunders. Their company, H. C. Baxter and Bros., had built the first freezing plant designed solely for the production of frozen french-fried potatoes.

Saunders recalls that the Maine company's first attempt showed promise in 1946, but at first fell far short of what the company eventually was able to produce. "The next time we used fresh potatoes instead of those that had been stored for a time and we were able to get better flavor and texture," says Saunders, who had been a USDA resident inspector at the plant. "The product was so light, though, it was hard to brown. John Baxter resolved that problem by dipping the fries in a dextrose solution, mainly corn syrup, covering the surface cells of the potatoes."

The U.S. production of frozen potato products surpassed 2 billion pounds in the early 1970s. The boom in the 1970s of irrigated acres was paralleled by increased plantings of Russet-Burbank potatoes. This variety is superior for french fry production and grows best in the Northwest states. That explains why, even in a rapidly growing market, the Northwest continued to produce around 80 percent of the nation's total annual output of frozen potato products.

As the 1980s began, 85 percent of the frozen potatoes packed in the Northwest were french fries. Most of the fries were purchased by the fast food industry. (p. 90)

The total annual per capita consumption of potatoes in the U.S. increased from 53 kg (117 pounds) in 1969 to 58 kg (127 pounds) in 1988 (Potato Statistical Yearbook, 1990). The amount of fresh potatoes consumed decreased from 28 kg (61 pounds) in 1969 to about 23 kg (50 pounds) in 1988. The increase in total per capita consumption is a result of increased consumption of frozen products, from 11 kg (25 pounds) in 1969 to over 20 kg (44 pounds) in 1988. This trend is anticipated to continue and indeed is even more obvious in export markets where the export of frozen potatoes, primarily frozen french fries, has tripled since 1980 (Table 3). Exports into developed and developing cultures are occurring primarily as a result of westernization of eating habits, the introduction of fast food restaurants, and increases in eating out in these countries (Table 4).

Potato imports have increased steadily over the past decade (Table 5). The most significant increase has been in frozen products, while seed imports have decreased.

As the world opens its markets to free trade, and tariff and quota barriers are removed, more potatoes and potato products will be traded among countries. Some regions of the U.S. may be adversely affected by these changes while others will prosper. The development of efficient production systems must be integrated with well-thought out and innovative marketing strategies in order for producers to remain competitive. Achieving market niches and providing unique products may be the key ingredients to future sustainability.

COMPOSITION, NUTRITIONAL VALUE, AND UTILIZATION OF POTATO TUBERS

The relatively high carbohydrate and low fat content of the potato makes it an excellent energy source for humans as well as livestock. In the United States, potatoes are, however, almost entirely used for human consumption. Only 1.3 percent is used for feed or starch and flour production (Table 6).

Since the potato produces large yields per unit land area, it is attractive as a major food source in areas throughout the world with high population density where protein or calories are limiting in the

TABLE 3. U.S. Potato Exports for Crop Years 1980-81 to 1987-88 by Usage.

Crop Year[1]	Table (MT)[3]	Seed (MT)	Dehydrated (MT)	Frozen (MT)	Total[2]
1980-81	116,701	8,798	34,422	44,282	473,439
1981-82	95,392	10,835	34,452	48,299	478,441
1982-83	83,505	4,801	24,625	52,061	389,428
1983-84	63,955	4,399	26,448	64,638	409,214
1984-85	44,404	4,676	19,632	64,253	334,642
1985-86	34,044	6,424	19,232	79,589	353,502
1986-87	44,017	5,675	27,516	105,410	480,640
1987-88	30,749	4,471	31,462	134,426	555,768

[1]Oct-Sept.
[2]Fresh-weight equivalent. Frozen conversion factor equals 2, and dehydrated 8.0.
Source: "Potato Facts" Fall/Winter, Bureau of Census.

(From 1990 NPC Statistical Yearbook, Page 49.)
[3]MT × 0.907 = U.S. ton

TABLE 4. United States Exports 1988-1989 by Destination

Commodity	Quantity (metric tons)^x 1988	1989	Value (U.S. dollars) 1988	1989
Tablestock Potatoes				
Africa	–	–	16,026	27,676
Canada	29,000	78,000	11,953,622	32,561,841
Central America	–	–	16,139	31,153
Caribbean	1,000	1,000	459,198	633,479
EC-Twelve				
West Europe	—	—	8,519	48,250
East Asia & Pacific	—	2,000	154,304	454,835
Middle East	—	—	—	7,812
Other	1,000	4,000	277,160	1,192,803
World Total	31,000	85,000	12,884,968	34,964,249
Seed Potatoes				
Canada	3,647	14,971	466,178	2,130,025
Caribbean	165	185	68,336	103,377
EC-Twelve	18	—	9,895	—
East Asia & Pacific	159	296	96,270	120,870
Other	82	1,553	12,848	375,431
World Total	4,071	17,005	653,527	2,729,703
Frozen French Fries				
Africa	—	133	—	97,111
Canada	370	493	315,498	368,545
Caribbean	1,203	1,863	687,062	1,504,033
Central America	9	452	11,434	296,057
EC-Twelve	588	83	464,105	114,253
East Asia & Pacific	104,582	132,825	68,609,940	88,548,248
Middle East	1,466	1,859	1,453,208	1,739,313
Other	45	70	27,314	43,007
World Total	108,263	137,778	72,568,561	92,710,567
Other Frozen Potatoes				
Canada	3,794	2,772	2,374,631	1,724,939
Caribbean	6,644	6,127	5,061,161	5,865,268
Central America	11	31	4,125	20,418
EC-Twelve	206	199	238,225	250,763
West Europe	93	15	93,420	21,324
East Asia & Pacific	7,080	10,480	5,910,000	8,040,259
Middle East	1	1	1,550	16,374
Other	68	404	49,289	217,199
World Total	17,897	20,039	13,732,401	16,156,544

TABLE 4 (continued)

Commodity	Quantity (metric tons)^x		Value (U.S. dollars)	
	1988	1989	1988	1989
Dehydrated Potatoes				
Africa	11	55	12,750	29,470
Canada	3,733	2,721	1,968,875	1,495,937
Caribbean	25	195	20,048	113,747
Central America	4	2	2,250	1,670
EC-Twelve	79	132	122,578	112,032
West Europe	152	149	178,312	221,584
East Asia & Pacific	1,014	691	942,284	707,191
Middle East	70	62	128,850	46,997
Other	82	68	42,693	44,678
World Total	5,170	4,075	3,418,640	2,773,306
Potato Flakes & Granules				
Africa	316	128	359,378	177,414
Canada	1,271	1,272	1,210,157	942,783
Caribbean	959	1,633	1,746,815	2,644,640
Central America	91	146	77,925	130,119
EC-Twelve	5,205	1,880	5,468,443	2,005,098
West Europe	900	1,281	1,291,568	1,324,941
East Asia & Pacific	16,153	19,919	10,617,190	14,600,403
Middle East	302	294	456,571	164,723
Other	843	368	758,324	384,810
World Total	26,040	26,920	21,986,371	22,374,931

Jul-Jun marketing year.
Source: USDA Foreign Agricultural Service.
(From: 1990 NPC Statistical Yearbook, Page 48.)
x: Metric tons $\times$ 1.102 = English tons

diet. Potato protein is considered to be of high biological value, although it is present in relatively low concentration (1.0-1.5 percent of the fresh weight). Potatoes also contain significant amounts of iron, thiamine, nicotinic acid, and riboflavin.

The primary components of potatoes are water (75-80 percent), starch (20-23 percent), protein (1.0-1.5 percent), and dissolved solids (3 percent). The relative importance of each of these components depends on the use of the crop. For example, potatoes used for home

TABLE 5. U.S. Potato Imports for Crop Years 1980-81 to 1987-88 (Metric Tons).

Crop Year[1]	Table[x]	Seed	Dehydrated	Frozen	Total[2]
1980-81	104,830	71,091	1,990	5,829	203,499
1981-82	163,525	61,595	1,450	8,642	254,004
1982-83	125,816	32,375	1,271	12,688	193,735
1983-84	94,949	32,075	2,700	19,605	188,834
1984-85	162,722	48,661	3,455	29,483	297,989
1985-86	106,294	28,041	3,647	34,513	232,537
1986-87	182,520	43,127	3,568	34,509	323,209
1987-88	175,450	42,252	3,292	49,796	343,630

[x]Metric tons × 1.102 = English tons
[1]Oct-Sept.
[2]Fresh-weight equivalent Frozen conversion factor equals 2, and dehydrated 8.0.
Source: "Potato Facts" Fall/Winter, Bureau of Census.
(From: 1990 NPC Statistical Yearbook, Page 49).

15

TABLE 6. Percent Utilization by Potato Product for 1983-1989.

Item	1983	1984	1985	1986	1987	1988[1]	1989[2]
Tablestock	32.1	31.3	30.7	30.2	33.2	31.1	31.1
Chips	13.0	11.7	10.4	12.7	10.4	12.1	12.1
Dehydrated	8.0	7.7	7.4	7.9	7.9	8.1	8.1
Frozen & French Fries	28.2	29.8	27.7	30.9	30.7	31.6	31.6
Canned	1.2	1.2	1.1	1.1	1.2	1.3	1.4
Total Food	82.5	81.7	77.4	82.8	83.4	84.1	84.2
Starch, Flour	.9	.9	.9	.8	.6	.4	.4
Feed	1.1	1.3	2.0	1.1	1.0	.9	.9
Seed	7.6	6.3	4.9	5.7	5.4	5.8	5.8
Total Other Uses	9.7	8.5	7.7	7.6	6.9	7.2	7.2
Shrinkage/Loss/Home Use	7.8	9.9	15.0	9.6	9.7	8.7	8.7
Total Production	100.0	100.0	100.0	100.0	100.0	100.0	100.0

[1]Revised.
[2]Projected. Proportionated to 1988.
Source: "Potato Facts," Fall/Winter, Agricultural Statistics Board, NASS, USDA.
(From: 1990 Potato Statistical Yearbook, Page 45).

consumption need to have a composition suitable for multiple uses, while those used for starch or dehydration must have a high starch content. In order to determine the appropriate variety to grow, where to grow the crop, what cultural practices to use, and which harvesting technique and storage procedures to use, a manager must have a good understanding of the composition of the potato tuber and the needs of the intended market.

Potato starch is composed of two types of molecules, amylose and amylopectin. These molecules are composed of individual sugar (glucose) molecules arranged in a specific pattern. Amylose is a straight chain and amylopectin is a branched molecule of individual glucose units. These components are made and degraded by specific enzymes and their ratio in the starch grain gives potato starch its characteristic cooking qualities.

Protein is an important component of potato tubers, even though it is present in relatively low concentration. Potato proteins are high in nutritive value, but more importantly are involved in developmental processes that affect the production of potatoes. Proteins function in diverse areas such as plant protection (proteinase inhibitors) and tuberization (patatin), as well as having many roles as metabolic pathway enzymes. The quality of the protein does not appear to vary significantly from one cultivar to another, even though the total quantity might. The two amino acids in potatoes that are limiting from a dietary standpoint are methionine and isoleucine (Hoff, Lockham, and Erikson, 1978) (Table 7). Cultural practices do not appear to affect protein quality significantly, although it is influenced by tuber maturity (Smith, 1968).

TABLE 7. Essential amino acid composition of proteins of three vegetable and two animal sources.

Protein	gms per 100 gms protein[x]				
	Potato	Corn	Wheat	Casein	Whole Egg
Arginine	5.0	4.8	4.2	4.2	6.5
Histidine	2.2	2.5	2.1	3.2	2.1
Isoleucine	3.7	6.4	4.2	7.5	5.7
Leucine	9.6	15.0	6.6	10.0	8.8
Lysine	8.3	2.3	2.7	8.5	7.2
Methionine	2.5	3.1	1.4	3.5	3.9
Phenylalanine	5.8	5.0	4.9	6.3	5.9
Threonine	6.9	3.7	2.9	4.5	5.3
Valine	5.3	5.3	4.3	7.7	8.8
Tryptophan	2.1	0.6	1.2	1.3	1.3
Tyrosine	6.0	6.0	4.0	6.4	3.6

From: Hoff et al., 1978
x: $9/100g \times 0.16 = oz/lb.$

Chapter 2

Potato Breeding

INTRODUCTION

New potato cultivars are developed to reduce problems associated with diseases and pests, meet new marketing demands, or to make general economic progress. Breeding potatoes to make these changes has not been as successful as it has been in other crops. The reasons for the lack of success are related to the genetic complexity of the potato.

Cultivated potatoes contain two (2x) to five (5x) sets of chromosomes with four sets (tetraploid (4x)) being common (previously shown in Table 1). Wild potato species may contain up to six sets of chromosomes. The high degree of ploidy (sets of chromosomes) in this crop makes it genetically diverse, but also makes it difficult to create new cultivars. When two parents with desirable characteristics are crossed, the recombination of chromosomes results in offspring that have both desirable and undesirable characteristics. As a matter of fact, most of the offspring (genotypes) that result are not significantly better than either of the parents. Therefore, conventional potato breeding projects have relied on large numbers of crosses and tests of large numbers of offspring to find the small number that are better than the parents.

The potato breeding and testing programs in the United States are located in diverse geographic areas and serve multiple market interests. Breeding programs are conducted by the United States Department of Agriculture (USDA) or land grant university personnel. The USDA has programs at Beltsville, Maryland, Aberdeen, Idaho, and Prosser, Washington. Land grant universities with active potato breeding programs are located in Colorado, Louisiana, Maine, Michigan, Minnesota, Nebraska, New York, North Dakota, Texas, and Wisconsin. The USDA plant introduction station at Sturgeon Bay, Wisconsin provides wild species materials to breeding programs throughout the United States. This station cooperates with international organizations such as CIP (the International Potato Center) in Lima, Peru to assure that genetic resources are available for basic potato improvement.

BREEDING OBJECTIVES

The major objectives of the potato breeding programs in the United States are quite similar. They are (1) yield enhancement, (2) improved raw product quality, (3) pest resistance, (4) improved storage characteristics, and (5) improved processing characteristics. These and other related breeding objectives are discussed in this section.

Yield

The yield component of potatoes is a complex character that is controlled genetically but modified to a considerable degree by management and the environment. Yield can be described as total weight of biomass per unit of land area, as the amount of output (value) relative to the cost of inputs (economic yield), or as the amount of energy required to produce the crop compared to that contained within the crop (energetic yield). Depending on which definition of yield is adopted, a breeder will select parents for crossing based on different characteristics. A large biomass yield may be beneficial for use in alcohol fuel production but uneconomical from the point of return in a fresh market production scheme. Some cultivars may have high yields only as a result of high input levels. Others may require more energy input than can be tolerated relative to the energy value of the crop produced.

Generally, each breeding project has selected a yield goal based on the yield of the most common cultivar grown at that location. For instance, in the western U.S., Russet Burbank is used as the standard for yield as well as for other characteristics. Although many other cultivars may yield more per acre, the average yield for Russet

Burbank is utilized because of its widespread use by the industry. In New York, cultivars such as Superior and Katahdin are used as fresh market standards whereas Monona is used as a processing cultivar standard. Yield targets utilized for dryland North Dakota conditions 114-136 kg (250-300 cwt [hundredweight]/acre) would not be suitable for areas with different climates or when potatoes are produced with irrigation.

The cultivar with the largest total yield may not be the most economic cultivar to grow for several reasons. The date at which a satisfactory yield is achieved is also important. Some evaluation trials harvest plots at two or three dates during the season in order to determine if a cultivar is best utilized as an early or a late cultivar. Fertility practices can modify the results and should be considered when planning this type of evaluation (discussed in Chapter 4). The market standards and subsequent value attached to quality factors such as grade, appearance, tuber size, and defects are a major consideration in selecting the correct cultivar. A high percentage (80-90 percent) of Number 1 USDA grade tubers is a part of the yield target of most improvement programs.

Skin Color

The skin color is important especially for selection of fresh market cultivars. Potato breeders in the North Central and Northeast may select round, white– or red-skinned potatoes whereas russet-skinned cultivars are preferred in the Pacific Northwest states. While the skin color may not seem important for processed potatoes, some considerations are commonly used. Russet-skinned potatoes are commonly selected in the West for processing into french fries because they are also in high demand by the alternative fresh market. Light skin color and/or a thin skin are desirable for processing so that the peel may be easily removed. Some specialty processed products may require that the skin be left on, which then may also dictate a specific skin type.

Shape

The shape of the tuber is also specific for different market needs. Round white potatoes are used for chips, boiling, canning, and dehydrated products, as well as in the fresh market. Oblong russet tubers are desirable for fresh market and also for processing as french fries. Oblong white or buff-colored potatoes such as Shepody are used in the processing industry as early harvest cultivars. Although Shepody can be used effectively as a main crop processing type potato, its fresh market prospects are low. Red-skinned potatoes, generally round, are used in the fresh market but can also be fresh processed if their composition is suitable.

Internal Characteristics

Internal qualities or compositional characteristics are of major concern. Some of the important characteristics are: (1) freedom from internal defects such as hollow heart (HH), vascular discoloration, brown center (BC), or internal brown spots (IBS); (2) flesh color; (3) flesh flavor; (4) dry matter and starch content; (5) sugar content; and (6) blackspot bruise susceptibility. These internal qualities are often modified considerably by the growing environment. Therefore, cultivar selection must be performed in the area in which commercial production will occur and tested over several growing seasons. It is commonly understood that hollow heart, brown center, and internal brown spots occur some years and not others, and that blackspot bruising is more serious some years than others. The specific gravity (related to starch content) of the tubers is also affected by growing season temperatures.

Plant breeders select clonal material that produces tubers which are free of internal defects by growing the genetic material in several locations for three to five years. This amount of time and number of locations is required to identify material that has resistance to internal disorders. Other cosmetic defects that may also be observed are elephant hide, growth cracks, and knobby or irregular growth. Although stressing plants for moisture and/or fertilizer may influence the amount of these disorders which occur, no routine screening procedures are commonly used (discussed in Chapter 8).

The carbohydrate (starch and sugar) composition is a very important character for both fresh and processing potatoes. In the case of fresh potatoes, high sugar content may be desirable for early fresh market potatoes to provide the sweet flavor of "new" potatoes.

High starch content is necessary to produce high quality "mealy" baked potatoes and french fries. Processing potato cultivars should have a low sucrose rating and less than 1 percent reducing sugars at harvest, and store well without sugars accumulating. For screening cultivars with the tendency toward high sugar accumulation, potatoes are stored at 5°C (42°F) for several months and then processed as chips or french fries. Reducing sugars may also be quantified by chromatographic or spectrophotometric techniques. The cultivars that do not fry brown are considered acceptable.

The total starch content is measured as specific gravity of a sample of tubers. Each cultivar can be rapidly and nondestructively measured by this technique (discussed in Chapter 10). The average specific gravity and the range of specific gravities within a lot of potatoes both need to be evaluated because their importance will be rated differently by end users.

Flesh color is variable in potatoes, although the most common cultivars have white or yellow flesh. Cultivars are being produced which have orange, purple, or purple and white flesh. Tubers with different flesh colors have different market niches. Certain flavors may also be associated with flesh colors, and taste panel results should be considered at some stage of the evaluation process.

Herbicide Susceptibility

Some cultivars have been found to be susceptible to herbicides that are registered for use on potatoes, and therefore trials may include screening for herbicide susceptibility.

Dormancy

Some breeding programs are now selecting genetic material and cultivars for increased tuber dormancy. If successful, increased tuber dormancy will decrease the need for the use of chemical sprout inhibitors. Selection for long dormancy must be moderated by considering the repercussions on seed germination. Tuber dormancy is also influenced by the growing season and affected by cultural practices such as planting date and fertility.

Growth Habit

The growth habit of the plant may influence other characteristics such as yield and disease resistance. If a plant type with excessive foliage is well-correlated with low yield and late maturity, selection against this characteristic may be needed. Dense foliage may contribute to susceptibility to foliar pathogens. If this is found to be true, then appropriate selection criteria can be developed.

Disease Resistance

Selection of potatoes for disease resistance has been a major concern of potato breeding projects at least since the potato famines of the mid-1800s. The late blight organism, *Phytophthora infestans*, is still a problem in potato production as are many other diseases. The common diseases that are screened for are Verticillium wilt, scab, late blight, early blight, and viruses such as potato leafroll virus (PLRV) and potato viruses PVA, PVX, PVY, PVS, and PVM. Other pests for which resistance may be sought are silver scurf; soft and dry rots; bacterial wilt; rhizoctonia; and cyst, golden, and root-knot nematodes. Each of these pathogens are explained in Chapter 7.

Screening for resistance to pathogens requires planting the selections in areas where the pathogen is naturally present or in areas that have been artificially infested. Oftentimes, a screening trial conducted early in the selection process may result in "escapes" because the field used for the screening is not uniformly infected. Unless resistance to the pathogen is one of the highest priorities of the program, it may be better to screen for it later in the program when enough seed is available to plant replicated trials.

Insect Resistance

Resistance to major insect pests has been difficult to achieve in potatoes. Most U.S. breeding projects screen for resistance to Colorado potato beetle feeding damage, and most screen for resistance to the leafroll virus transmitted by the green peach aphid. These trials are generally conducted in a manner similar to those used for disease resistance trials. Plants are grown under field conditions which are conducive to the pest and no control sprays are used. Resistance to the Colorado potato beetle is obvious because when the insects are present, resistant plants are not consumed rapidly or as completely as susceptible plants. Full resistance has not been found in cultivars having acceptable horticultural characteristics at this time. Resistance to potato leafroll virus (PLRV) needs to be confirmed by tissue testing for presence of the virus. Screening is also conducted for the potato leafhopper in some states.

Hierarchy of Breeding Objectives

The parameters that are commonly used in the breeding programs are shown in Figure 2. Each individual characteristic or group of characteristics must be prioritized so that a hierarchical system can be developed. The screening procedures for diseases, insects, and herbicides can be subdivided for those important in each breeding program.

BREEDING METHODS

The primary method for obtaining new cultivars in the United States is by the method of recurrent selection. With this method, cultivars, clones, or lines that are to be used as parents are chosen because they have the desired characteristics. These are crossed by removing the anthers from one parent (female) and transferring the pollen from the other parent (male) to the stigma of the female plant (Figure 3). Successful pollinations result in seeds being produced in small fruit (Figure 4). Many cultivars such as Russet Burbank do not produce seeds or fruit readily and therefore make poor parents. Following pollination, the seeds are allowed to develop to maturity. They are harvested and planted and the resulting seedlings are tested for the presence of the desired characteristics by the appropriate screening techniques. The best progeny are selected and the process is repeated using the new progeny as parents. Sometimes the offspring are combined with one or both of the original parents (backcrossing).

The selection of parents is crucial to the breeding process and key characteristics must be identified in the parents in order to achieve the improvement desired. Some programs, such as in New York for exam-

FIGURE 2. A breeding system hierarchy.

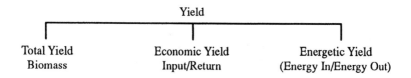

Yield

Total Yield   Economic Yield   Energetic Yield
Biomass   Input/Return   (Energy In/Energy Out)

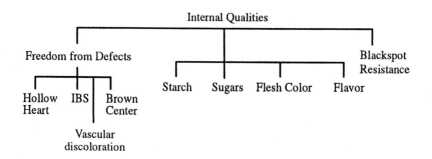

Internal Qualities

Freedom from Defects   Blackspot
                                                       Resistance
Hollow   IBS   Brown   Starch   Sugars   Flesh Color   Flavor
Heart   Center

                Vascular
                discoloration

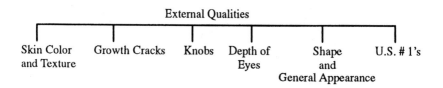

External Qualities

Skin Color   Growth Cracks   Knobs   Depth of   Shape   U.S. # 1's
and Texture   Eyes   and
                                              General Appearance

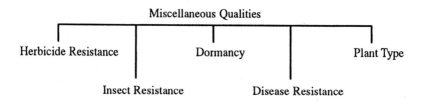

Miscellaneous Qualities

Herbicide Resistance   Dormancy   Plant Type

          Insect Resistance   Disease Resistance

FIGURE 3. A typical potato flower.

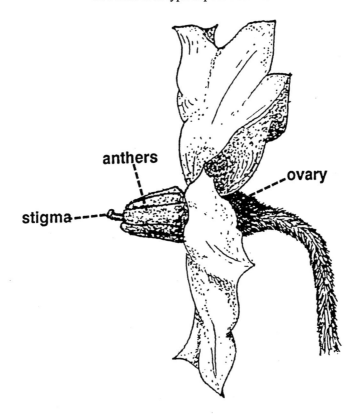

ple, select parents which are both good pollen sources and resistant to golden nematode and scab in order to produce a potato acceptable for processing into chips. The number of offspring produced each year varies with the size of the program and its goals. For example, with the USDA program in Aberdeen, Idaho, enough crosses are made so that 100,000 seedling plants can be planted and evaluated each year. Between 100 and 150 parents of both 4x and 2x genotypes are utilized for crossing.

Most breeding programs follow similar protocol for evaluating new genetic materials. The first-year seeds, seedlings, or tubers from greenhouse-grown plants are planted in the field to determine the general tuber type and relative yield (year 1). Individual tubers or seeds are often distributed to other breeding and testing programs at this stage. The second year's trials consist of 2,000-3,000 12-hill plots (one hill = one plant) which are again evaluated for tuber type, relative yield, tuber specific gravity, frying quality after being stored at 7.5°C (45°F) and 5°C (42°F), and general storability (year 2). Replicated trials may be planted during the third year at two or more locations depending on the availability of seed (year 3). The following year is used to screen for disease resistance (year 4) followed by distribution to other evaluators in the fifth year. During the next two years (years 5 and 6), the cultivar will be evaluated for overall acceptability and may be included in regional cultivar trials. Cultivar trials are conducted by state university faculty who may be horticulturists, pathologists, or from other disciplines and not necessarily plant breeders. They bring varying experience to the procedures which are key to the successful evaluation of new cultivars. After two or three years in regional trials, a clone will be considered for release as a new cultivar (years 7 to 9). At this time, seed may be increased to commercial grower lot size, which will take one to three more years (years 10 to 12). If the cultivar has continued to show advantages over the current cultivars, full scale commercial evaluations will be made at this time. It may take two to four years for fresh packers or processors to become familiar enough with the cultivar to determine if it will be a successful cultivar (years 12 to 15).

Developing new cultivars is a long-term commitment to methodical evaluation of a multitude of traits. The outcome of the process in the United States has been slow but is now beginning to show benefits of the laborious efforts of potato improvement programs. It is crucial that breeders establish screening protocol and develop a hierarchical structure in their programs if they are going to be successful.

Relatively new techniques (biotechnological procedures) have been implemented in some breeding programs with the hope of speeding the process of cultivar improvement. These techniques include isolating genes (the genetic codes) responsible for specific traits and incorporating the gene(s) into existing cultivars. For example, the gene for production of the toxin from *Bacillus thuringinsis*, which kills Colorado potato beetles, has been isolated, cloned (reproduced), and inserted

FIGURE 4. A potato fruit or seed ball showing developing seeds.

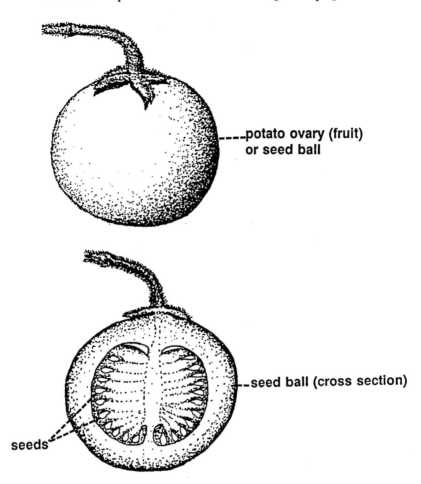

into potato cells. This is a complex process that may result in plants that are not damaged significantly by the Colorado potato beetle. At this time, results are too preliminary to determine the potential success of this procedure. Other examples include incorporation of the genes which provide resistance to PVX, PVY, and PLRV, and incorporation of glyphosate resistance (herbicide resistance) into tomatoes. The future of these techniques is exciting but uncertain.

The selection of cultivars for growing in specific environments has been largely empirical. A cultivar is planted and observed for several years, perhaps with varying fertility or irrigation practices, and its overall suitability is evaluated. This procedure has resulted in the establishment of cultivars suited for fresh market use and others selected for processing.

Fresh market cultivars must produce tubers of relatively uniform size 114-340g (4-12 ounces), of smooth appearance with few internal or external defects, and be resistant to disease. Cultivars may have a red, white, or russet skin and be either round or oblong in shape. The internal qualities or cooking characteristics may not be the same for fresh market cultivars as those required for processing varieties since fresh market potatoes have a multiplicity of uses by the consumer. Flavor is of concern particularly if the cultivars are also to be grown in the home garden. It is desirable for fresh market cultivars to have a relatively long tuber dormancy after harvest so they can be stored for long periods, by the retailer and the consumer, under conditions that are not optimum.

Tubers of processing cultivars need to meet specific requirements for cooking, depending on the processed product being made. For frozen fried products, they should process to a mealy texture (fluffy and not hard or soggy), be low in reducing sugars to prevent browning when fried, and the skins should be easy to remove. Elongated, block-shaped tubers provide the best recovery for french-fried products. Russet-skinned types have been used for processing in many instances so that when market demands provide an economic incentive, they can be used for the fresh market as well. Potatoes for chip processing should have a specific gravity (discussed in Chapter 10) of 1.080 to 1.095, must fry with acceptable light color, and be free of defects. Round tuber types are most desirable for chip processing. Processing cultivars should have high starch contents, especially if used to produce dehydrated potato products.

CULTIVAR CHARACTERISTICS

Many potato cultivars are available to producers. Specialty potatoes with yellow flesh or purple skin, heirloom varieties, and new releases are all found within the industry. Producers should consult seed production catalogs for availability of cultivars to test. The fourteen cultivars which occupy the largest amount of production are discussed here and reported in Table 8.

Atlantic: This is a processing potato used in the chipping industry. It has high specific gravity and high yield potential. Tubers are round to oval, skin is netted with a scaly appearance, and they have white flesh. It is susceptible to scab; somewhat tolerant to Verticillium wilt; resistant to pinkeye; and highly resistant to race A of golden nematode, virus X, and net necrosis from PLRV (*North American Potato Varieties Handbook*, 1990). The tubers are also susceptible to a physiological disorder called heat necrosis or internal brown spot (IBS), which results in necrotic regions near or in the vascular tissue. It is also susceptible to hollow heart and may have poor storage qualities.

TABLE 8. Fourteen major potato cultivars which passed certification in the United States, 1989.*

Cultivar	Date Introduced	% of Seed Acreage (last 5 years)
Russet Burbank	Before 1914	38.0
Norchip	1968	8.4
Atlantic	1976	6.1
Russet Norkotah	1987	5.9
Superior	1961	4.5
Kennebec	1948	3.6
Norland	1947	3.7
Centennial Russet	1976	3.3
Shepody	1980	2.0
Norwis	1990	-
Monona	1964	2.5
Red Pontiac	1949	2.4
Red LaSoda	1953	1.9
Katahdin	1932	1.9

* Source: National Potato Council Potato Statistics Yearbook. 1989.
  Dr. R. W. Chase, Michigan State University.

Centennial Russet: Centennial Russet is a fresh market potato grown largely in Colorado and California. It has average qualities and yield but is not processed because of its high sugar content. The tubers are oblong, smooth, blocky, and slightly flattened with a thick, dark-netted skin (*North American Potato Varieties Handbook*, 1990). It has moderate tolerance to Verticillium wilt, early blight, Rhizoctonia, and Fusarium dry rot, but is susceptible to common scab and PLRV.

Century Russet (A74212-1): This new cultivar is used as a fresh market potato with good baking and boiling characteristics but does not process well. It has a high yield, moderate specific gravity, white flesh, and a light russet skin. Century Russet is highly resistant to Verticillium wilt and PVX and moderately resistant to common scab and foliar early blight, although tubers are susceptible to early blight and Fusarium dry rot (Rykbost, Carbon, and Voss, 1990). It is resistant to internal disorders, but very susceptible to harvest damage and associated diseases.

Katahdin: Katahdin is a fresh market cultivar in the Northeast, but it has significantly decreased in importance since the 1970s. Its round to oblong tubers have white (buff) skin and creamy white flesh. The specific gravity and cooking quality are moderate. It is susceptible to scab and resistant to mild mosaic, net necrosis and brown rot (*North American Potato Varieties Handbook*, 1990).

Kennebec: This cultivar has been used primarily by the processing industry. It produces large yields of oblong to round tubers that may show significant amounts of hollow heart. The specific gravity is moderate, and the skin color and flesh are white to cream color. It is moderately resistant to late blight, is resistant to net necrosis, and is very susceptible to Verticillium wilt (*North American Potato Varieties Handbook*, 1990). Sugars build up in this variety if stored below 10°C (50°F).

Monona: Monona is used as a chipping cultivar in the Northeast and North Central regions of the United States. It has oblong to oval tubers with a light cream (buff) skin and white flesh. Monona stores well at 10°C (50°F) and reconditions rapidly from cold storage. It is resistant to mild and rugose mosaic, tolerant to scab and Verticillium wilt, and is very susceptible to blackleg (*North American Potato Varieties Handbook*, 1990).

Norchip: This is a white-skinned, round, high specific gravity chipping cultivar. The flesh color is white, and yields may be low. It is tolerant to common scab and tuber flea beetle, but is susceptible to late blight, Verticillium wilt, PVX, PLRV, spindle tuber, and early blight, and to sugar buildup if stored below 10°C (50°F) (*North American Potato Varieties Handbook*, 1990).

Red LaSoda: This cultivar is used for the fresh market primarily in the Southeast, northern California and southern Oregon. It develops good red color in the Southeast, but only moderate in the West. The color fades during storage. The flesh is white, and specific gravity is usually low. It tends to produce large tubers with deep eyes, and may be irregular in shape. Red LaSoda is susceptible to blackleg, PVX, PVY, PLRV, early and late blights, scab, corky ring spot, and Fusarium and bacterial wilts (*North American Potato Varieties Handbook*, 1990; Rykbost, Carbon, and Voss, 1990).

Red Pontiac: Red Pontiac is also known as Dakota Chief. It is a fresh market potato with round to oblong deep eyes, red-colored skin, and white flesh. Red Pontiac is susceptible to most common potato diseases (*North American Potato Varieties Handbook*, 1990). Tubers may become oversized and irregular in shape, and high yields are obtainable. It is resistant to after-cooking darkening.

Russet Burbank (Idaho Russet, Netted Gem): This is the most prominent potato cultivar grown in the United States. It accounts for 38 percent of the potato seed acreage and is the major cultivar used for potato processing in the Northwest and North Central regions. Tubers are oblong and slightly flattened with a russet skin, white flesh, and high specific gravity. It can be used for both fresh market and processing because of its culinary qualities. Russet Burbank is susceptible to producing irregular-shaped tubers and having internal disorders such as hollow heart, heat necrosis, net necrosis, jelly end, high sugar context or sugar end sugars, and internal brown spot. It is tolerant to scab, but is susceptible to Fusarium and Verticillium wilts (*North American Potato Varieties Handbook*, 1990), late blight, PVX, PVY, PLRV, and white mold (Rykbost, Carbon, and Voss, 1990).

Russet Norkotah: This new, early, fresh market potato cultivar has an attractive dark russet skin, white flesh, and oblong tuber shape. It has low to moderate specific gravity and is resistant to most internal disorders. Susceptibility to diseases such as early blight, white mold, late blight, black dot, Verticillium wilt, and most viruses make performance inconsistent. It yields erratically, but has a high percentage of U.S. No. 1 tubers.

Sebago: Sebago is late-maturing and best suited for fresh market because of its good cooking qualities. Tubers are white-skinned with shallow eyes and medium specific gravity. It is resistant to mild mosaic, net necrosis, scab, and late blight. Commercial acreage has decreased substantially during the past decade.

Shepody: Shepody has become a significant processing cultivar for the early harvest season in the West, and out-of-storage in the eastern United States and Canada. It has long to oblong tubers, white to buff skin, and white flesh. The yields and specific gravities are moderate to good and it stores well. Shepody is moderately resistant to net necrosis, rhizoctonia, Fusarium, and early blight. It is susceptible to PVS, PVX, PVY, PLRV, pink eye, pink rot, Verticillium wilt, and late blight and is very susceptible to common scab and powdery scab (Rykbost, Carbon, and Voss, 1990). Its susceptibility to the herbicide metribuzin, is a detrimental characteristic, and it may form rough tubers when they become large.

Superior: Superior is the most popular round, white, early, fresh market cultivar. It has very good general appearance, round to oblong tubers with a buff-colored skin, and moderate yields. It is resistant to net necrosis and common scab, but is susceptible to blackleg, Fusarium, PLRV, PVX, PVY, and Verticillium wilt (*North American Potato Varieties Handbook*, 1990).

Chapter 3

Potato Seed

INTRODUCTION

The potato crop is one of the few annual crops planted using vegetative tissue (pieces of the potato tuber) instead of botanical seed. Although production of potatoes from botanical seed is entirely possible, commercially acceptable crops have not been produced at this time in the United States. The reason for the inability to produce commercial crops from botanical seed is genetic variability, which results in nonuniform tuber quality and substantially lower yields. If this problem is eventually overcome, the use of botanical seed for commercial production may come about. In some areas of the world where tuber uniformity is not an important quality consideration, production from botanical seed is practiced.

PRODUCTION REGIONS

Potato seed is usually produced in regions that are isolated from commercial potato production, with the intent of preventing disease contamination by insects, animals, or human traffic. In the West, most of the premium seed growing areas are located in short growing season mountain valleys, or plateaus where insect populations are reduced. The distance to commercial markets often makes these areas unattractive for other production schemes, so they are suited to be organized into seed production only (quarantine) areas. The suitability of these areas may also be based on their tendency to not support insects such as the green peach aphid (discussed in Chapter 6) during the fall, winter, and spring months. The general cultural requirements for seed growing are similar to the commercial crop, so only the unique aspects of potato seed production will be covered in this chapter.

CERTIFICATION

Certification of potato seed in the United States has been refined significantly during the past 20 years. Several states require that commercial growers plant only certified seed in order to reduce losses from ring rot, blackleg, viruses, nematodes, and other pathogens. The procedures used for certification vary from state to state and do not imply that seed purchased, following successful completion of the certification procedure, is free of disease. It does, however, provide the purchaser with a set of standard tests through which the seed has passed by various means of inspection. The following sections will illustrate this process.

Initiation of a seed production program first requires that the individual select an approved source of plant material. The plant material can be (1) tubers, stem cuttings, meristems (cuttings from the apex of virus free plants), or derived plants purchased from a commercial firm or grown by the individual seed grower; (2) tubers selected in the field by the grower (line selection); or (3) stock seed purchased from a commercial seed grower. The criteria for selecting the plant material will differ depending upon the requirements of the seed certifying agency and the level of sophistication of the potential seed grower.

A grower may make a selection of a single tuber or a small sample of tubers in a field based on the horticultural characteristics of the tuber such as shape, color, specific gravity, freedom from disease, etc. The explants produced from the selected seed tubers are tested by a reliable laboratory to determine if the plant is free from the predetermined diseases (discussed in Chapter 7). If the material is not disease-free, it may be cleaned by growing plants from the tubers at elevated temperatures, dissecting the shoot apex, and isolating the meristem. This meristem can be propagated aseptically and retested. Once the plant material has been determined to be free of disease, it is considered pre-nuclear and may be planted in the field. Most seed growing states have adopted a "limited generation" program for seed production (Table 9). The generation program identifies a starting point for each seed lot and gives it a generation designation. The first generation is usually designated nuclear and is the first generation raised from seed tubers or meristem plants in a field situation. The specific requirements for each generation vary from state to state, but are clearly defined in their certification rules and regulations. Each generation has specific testing requirements for disease content in order to be maintained as certified seed. These requirements all change periodically and, therefore, current certification standards should be obtained from each state from which seed is to be procured to compare seed testing procedures. An example of the disease tolerances set for each seed generation is given in Table 10.

These disease tolerances apply to PVX-tested, stem-cut, meristem, line-select, and non-PVX-tested classes. "Zero" tolerance–the 0.0 percent tolerance–is not intended, nor may it be construed, to mean that the lot inspected is free from the disease. In cases of bacterial ring rot and Columbia root-knot nematode, it means only that the stated disease was not found to be present during the inspection process. In the case of other diseases, it means only that the diseased plants observed were required to be rogued out.

Each state certifies the growers' seed lots by sampling for virus, testing, and visually inspecting the fields to determine the content of each disease that has been identified as important by the certification agency. Inspectors walk through representative areas of each field to obtain their sample count of the number of diseased plants. These numbers are recorded for each inspection. Inspections usually occur just prior to row closure, during midseason, and again late in the season. The number of inspections varies depending on the state, the condition of the field, and weather conditions (such as frost). In addition to the field tests, a storage bin inspection and winter grow-out tests may be performed. Storage inspections may be used to advise growers about the general appropriateness of storage or handling techniques. The winter test or alternate season tests used by some states provide a means to grow and evaluate the seed prior to it being accepted for certification. Results of the field inspections are published by the certification agencies and are avail-

TABLE 9. Seed Class Terminology Chart

Generation Terminology Used for Field Planting						
AGENCY	1st	2nd	3rd	4th	5th	6th
Alaska	G-1	G-2	G-3	G-4	G-5	G-6
California	N	G-1	G-2	G-3	F	C
Colorado	G-1	G-2	G-3	G-4	G-5	G-6
Idaho*	N	G-1	G-2	G-3	G-4	G-5 G-6
Maine**	Maine Potato B. Farm			G-1	G-2	G-3
Michigan	N	G-1	G-2	G-3	G-4	G-5
Minnesota	N	G-1	G-2	G-3	G-4	G-5
Montana	N	G-1	G-2	G-3	G-4	—
Nebraska	N	G-1	G-2	G-3	G-4	G-5
New York	Uihlein Farm		FU-1	FU-2	FU-3	F
North Dakota	N	G-1	G-2	G-3	G-4	G-5
Oregon	N	G-1	G-2	G-3	G-4	G-5
Utah	G-1	G-2	G-3	G-4	G-5	G-6
Washington	N	G-1	G-2	G-3	G-4	—
Wisconsin	U of W Farm		FG-1	FG-2	FG-3	FG-4
Canada	P-E	E-1	E-2	E-3	F	C

Abbreviations Used

G = Generation   E = Elite
N = Nuclear   P-E = Pre-Elite
F = Foundation   FU = Foundation Uihlein
C = Certified   FG = Foundation Generation

*Idaho has a G-6
**Maine has a G-4 and G-5
Clark, Richard S. 1990, for the Certification Section; Potato Association of America, revised April 1990.

able to seed purchasers. The storage and winter tests should be available directly from the seed grower. The names and addresses of the official certification agencies are included in the Appendix.

The certification results provide the buyer with a guide to the quality of the seed produced by seed growers in each state. It is difficult to compare the results between states because of differences in certification requirements. The buyer is advised to visit seed growers who are prospective seed suppliers, inspect their

TABLE 10. Maximum Field Disease Tolerance as Percent for Certification (Montana).

Disease	Generation II 1st, 2nd & 3rd Inspections	Generation III 1st, 2nd & 3rd Inspections	Generation IV 1st, 2nd & 3rd Inspections
Virus Leaf Roll	0.062	0.125	1.25
Mosaic	0.1	0.2	1.0
Spindle Tuber	0.0	0.0	0.0
Calico	0.15	0.3	1.0
Haywire	0.25	0.5	1.0
Witches Broom	0.25	0.5	1.0
Giant Hill	0.05	0.1	0.5
Blackleg	0.062	0.125	0.25
Varietal Mix	0.0	0.0	0.25
Bacterial Ring Rot	0.0	0.0	0.0
Columbia Root Knot Nematode (visible symptoms)	0.0	0.0	0.0

fields, handling, and transportation facilities, and review sanitation, pest control, and seed source selection practices. The development of a strong interaction between the seed grower and buyer is important to long-term success.

SEED CUTTING

Proper cutting of potato seed tubers is important if maximum productivity of the seed lot is to be realized. The primary outcomes desired of a good seed cutting operation are uniform-sized seed and minimum spread of disease. Seed should not be cut directly out of cold storage. The seed of some cultivars will rot and produce very poor stands if cut prior to a period of warming (the reasons for this are not known). Seed tubers should be stored with good ventilation at 10°-15°C (50°-60°F) for seven to ten days prior to cutting and planting.

Seed cutting is accomplished by passing potato tubers through rotating discs used as knives (Figure 5). The size of the cut seed can be changed by adjusting the sizing rollers to allow more or fewer tubers to pass through the top set of knives, which make a vertical

FIGURE 5. Potato seed cutter.

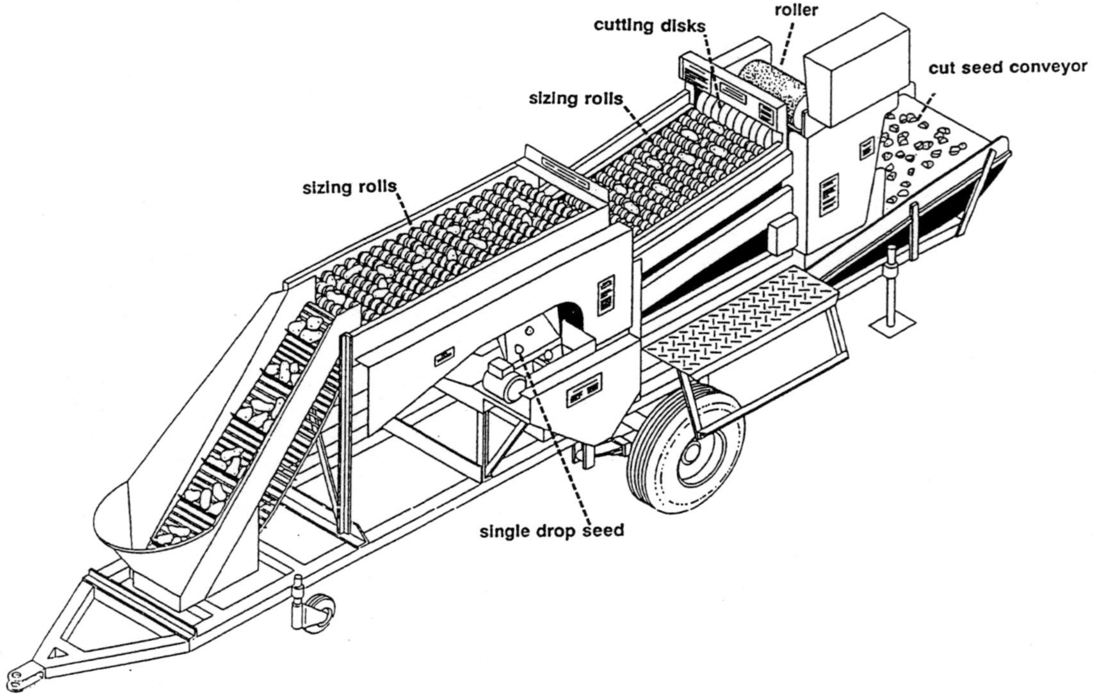

cut seed conveyor

roller

cutting disks

sizing rolls

sizing rolls

single drop seed

ESKEL seed cutter by Better Built Potato Seed Cutter Co.

as well as a horizontal cut, or through the second set of knives below, which make only a vertical cut and can be spaced to vary the distance between the vertical cuts. The appropriate control of flow through the cutter, allowing proper sizing, selection of the appropriate distance between knives, and maintenance of the rollers to give good horizontal cuts are critical for a good seed cutting job. An example of the profile of seed delivered from a seed cutter is illustrated in Table 11. Producers should invest in an accurate scale to weigh seed pieces to determine the size profile, or have the commercial seed cutter operator report cut seed profiles and average weights to assure that the seed is cut according to the prearranged specifications.

An approximation of the size of cut seed can be made by collecting samples of cut seed, counting the number of seed in a 4.5 kg (ten pound) sample, and comparing it with Table 12. This is only an estimate of average size and does not tell how much variation in size there is. Small pieces of seed (chips) and oversized pieces should be estimated in each sample because they cause reduced plant vigor or poor planter performance. Use the Seed Cutter Evaluation Worksheet found in the Appendix for checking seed cutter performance.

SEED TREATMENTS

Potato seed may be treated following cutting for one or two reasons. If the seed has been harvested and is then scheduled for planting before dormancy has been overcome, treatment with 1-2 parts per million (ppm) of gibberellic acid (GA) can be used to break dormancy. This practice has been recommended for the winter crop in Florida. Seed may also be treated in some growing regions for control of fungal or bacterial diseases. In these cases, the chemicals are usually formulated into a powdered inert material so that the seed may be dusted. Some chemicals have been used as a liquid spray, but danger lies in getting seed too wet which may lead to rot in the field. Producers should test each seed treatment separately in their own areas to determine if it is an effective input.

TABLE 11.. The percentage of seed pieces in different size clases from seed cutting operations.

	Percent of Seed by Size Class				Av. Seed weight (oz)	Av. Seed weight (oz)
	<28 gr <1 oz	28-42 gr 1-1.5 oz	42-56 gr 1.5-2.0 oz	>56 gr >2 oz		
1984 Av.	7	29	42	23	1.7	48
1983 Av.	10	24	35	32	1.8	50

From: G. Q Pelter. 1984. Columbia Basin Seed Size Survey. *Spud Topics*, Vol. 30 (21). April 24, 1985.

Current treatments may be recorded in the space below for quick reference.

Current seed treatment

a.
b.

Purchasing good quality seed, sanitizing equipment, cutting warmed seed, and planting when soil temperatures are above 10°C (50°F) are the most important factors in achieving the desired number of vigorous plants per acre in the fields.

Seed may be suberized, the wounded surface healed, following cutting by placing it at 15°-20°C (60°-68°F) and 85-100 percent relative humidity (rh) for a period of three to seven days to reduce seed piece breakdown. This procedure may be effective in reducing seed piece decay losses in some cultivars, especially when planting into soils at temperatures less than 10°C (50°F).

PLANTERS AND SEED PLACEMENT

Potato planters consist of three basic designs; the cup planter (Figure 6), the pick planter (Figure 7), and the assist-feed planter (not shown). The cup planter usually plants at a higher rate of speed than the pick-type, however, the pick planter usually places seed more accurately than the cup planter (Misener, 1982). Slower speed, single row cup planters may plant more accurately than the double cup style. The third type of planter, called an assist-feed planter, is accurate but not suitable for large acreages unless speed of planting is not critical.

The cup and pick planters are quite similar in design except for the mechanism that picks up the seed. The cup planter utilizes plastic or metal cups attached to a chain to convey seed from a delivery belt at the bottom of the storage hopper or bin. The cups transport the seed vertically upward, invert themselves at the top, and the seed falls on the cup ahead of it. The cup travels down the mechanism until it makes a second turn at the bottom of the planter, the seed falls into the planter shoe and is covered with soil by the covering disks. The cups should be sized for the seed size used by each producer. Seed that is too large for the cups may fall off and leave skips in the field. Seed that is too small may result in two or more seed pieces being placed together.

The pick planter differs by picking seed out of a storage compartment using sharp picks of various lengths. The picker arm is spring-loaded so that it forces the picks into seed pieces at the bottom of the seed storage compartment, carries the seed to the

Table 12. Number of seed potatoes in a 4.5 kg (10 pound) sample of seed.

Average Seed Size	# of Seed per 4.5 kg (10 lbs)
28 gr 1.00 ounce	160
35 gr 1.25 ounce	128
42 gr 1.50 ounce	107
49 gr 1.75 ounce	91
56 gr 2.00 ounce	80
63 gr 2.25 ounce	71

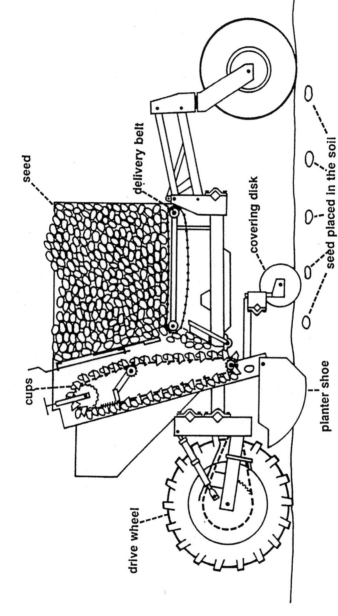

FIGURE 6. A cup potato seed planter.

Cup potato seed planter by Acme Manufacturing Co., Inc.

FIGURE 7. A pick potato seed planter.

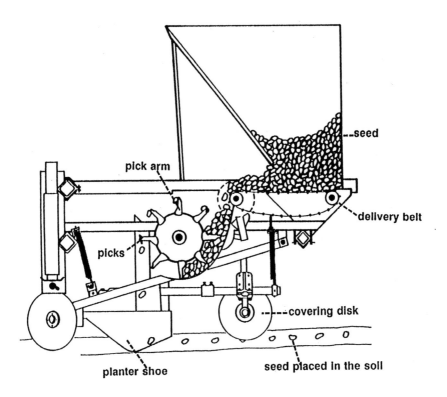

Pick planter by Dahlman Co.

forward position, and then releases the seed by retracting the picks when the two pieces of the picker arm are separated by the spring mechanism. The seed falls into the planter shoe and is covered by the covering disks. The choice of pick length must be consistent with the speed of operation. The picks should be replaced when they become bent or dull from damage by rocks or other debris. The configuration (short vs. long) and positions are determined by the owner's manual and performance evaluations.

Neither planter will place poorly-sized seed pieces properly in the field. Well-cut and sized seed can be placed accurately enough in the field by either planter if they are properly adjusted, maintained, and operated at appropriate speeds. The amount of seed being fed to either mechanism must be controlled to allow an even flow at an appropriate depth so that the cups or picks can function properly. Shakers may be adjusted on a cup planter to eliminate or reduce doubles. The manufacturer should supply adequate information to ensure proper adjustment of the planter. Adjustments should be checked regularly to ensure they are correct.

Seed placement can be determined only by carefully uncovering the seed following planting and measuring its placement. The accuracy of planting can be calculated in a number of ways and only one example is given here. Uncover 3 meters (ten feet) of each row across the width of the planter and measure the distances between each of the seed pieces. Since in-row spacing was determined by the producer ahead of time, use the chosen spacing to signify100 percent accuracy. For example, if the desired spacing was 23cm (nine inches) between plants, in each 3 meters (ten-foot) section there should be 13 seed pieces. If there are 13 seed pieces, then the planter is placing the correct number of seed per acre. The seed may not have been placed where you wanted them, however. Check to see how many seed pieces are less than 50 percent of the distance away from the previous seed (double planted), or how often there is no seed present at 150 percent of the distance from the previous seed (a skip). The number of doubles plus the number of skips, subtracted from the total number, and then divided by the total number equals the accuracy value. For example, if 13 seed pieces were found in ten feet of row but there was one double and two skips, then: 13 total − (1 double + 2 skips) ÷ 13 = 77% accuracy. Calculations of this type can be used to check new planting equipment, to evaluate new styles of planters and to adjust planters for optimum operation. Use the "Seed Placement Evaluation Worksheet" and "Seed Placement Form" found in the appendix for evaluating planter performance.

The accuracy of seed placement may affect the yield and quality of the crop. The economics of planter speed and the difference in yield between a well-adjusted cup planter and a comparably adjusted pick planter has not been easy to determine (Misener, 1982; Pascal, Robertson, and Longley, 1977). Therefore, each producer should evaluate the accuracy of the planter being used, and calculate the machine's value depending upon (1) its cost and (2) its ability to plant from either a speed or quality standpoint or both.

The optimum plant population for potatoes depends on the cultivar, soil conditions and fertility practices, the length of the growing season, water availability, method of irrigation, and intended market. Cultivars that produce large plants which mature late with high tuber yields will require more space than cultivars that produce smaller, short season plants. Plants grown under natural rainfall generally require wider spacing than those grown under irrigation. Svenson (1977) found that the tuber yield per stem was 1.7 times greater with plants placed 20 cm (8 inches) apart compared to those at 10 cm (4 inches) spacing, although yield per unit, land area was only 5 percent greater. The wider plant spacing also resulted in a greater number of tubers per stem and larger tubers. A larger amount of foliage per unit, land area was produced with closer, 10 cm (4 inch) spacing. Others have found that plant spacing from 22-30 cm (9-12 inches) results in the best yields (Bremner and Taha, 1966; Bushnell, 1930; Claypool and Morris, 1931). It has been suggested that there is no advantage to planting in beds rather than rows, and that higher populations tend to reduce the tuber size even though total yield might be increased. No studies have been reported that determine the interactions of spacing, fertility, and water application. Allen (1981) suggests that "it is also important to examine more closely the physiological responses which allow this crop to tolerate such extreme spatial arrangements."

Generally, the higher the plant populations, the larger the number of stems per acre, resulting in a larger number of tubers of smaller size. Therefore, the decision of plant density must be made in accordance with the ultimate use of the crop. For example, the most desirable potatoes for the fresh market may be smaller than those going to processing.

If the average seed piece size is known, the plant population or spacing can be checked by comparing the pounds of seed planted per acre with the following table (Table 13).

TABLE 13. Amounts of potato seed used per acre for different row spacing and seed size.

	Pounds of seed used per acre								
	Row spacing								
	32 inches			34 inches			36 inches		
	Inches between plants								
Seed Size (oz)	6	9	12	6	9	12	6	9	12
1.00	2047	1361	1020	1917	1278	962	1815	1210	908
1.25	2559	1701	1275	2396	1598	1203	2269	1513	1135
1.50	3071	2042	1530	2876	1917	1443	2723	1815	1362
1.75	3582	2382	1785	3355	2237	1684	3176	2118	1589
2.00	4094	2722	2040	3834	2556	1924	3630	2520	1816
2.25	4606	3062	2295	4313	2876	2165	4084	2723	2043

Pounds × 0.454 = kilograms   Ounces × 28.35 = grams
Inches × 2.54 = centimeters   Acres × 0.405 = hectare

Between-row planting distances are usually 80-90 cm (32-36 inches) with plants spaced 22-30 cm (9-12 inches) apart. Seed is placed 10-15 cm (4.5-6 inches) below tilled ground level and planted with a hilling or bedding operation. The single most important factor in favor of planting in hills is ease of digging. Furrows provide a stable guide for the harvester and less soil needs to be lifted at harvest when the tubers are in hills rather than in a flat arrangement. With the use of effective herbicides, bed and flat plantings are more feasible compared to when frequent cultivation was needed to control weeds. New harvester designs with reduced drag may make bed or flat field planting more widely used in the future.

PEST CONTROL

Pest control in seed potato fields is essential to the success of the industry. Insect pests such as the green peach aphid must be controlled to prevent disease in succeeding plantings. The specific means of insect control are covered in Chapter 6. However, it must be emphasized that insect control is one of the most important aspects of potato seed culture.

Specific disease control is covered in Chapter 7, but emphasis must be placed on specific issues related to seed at this time. The two bacterial pathogens which cause the most significant problems in the seed industry are the soft rot or blackleg organism, *Erwinia carotovora,* and the ring rot bacteria, *Clavibacter (Corynebacterium) michiganense* subsp. *sepedonicum.* Both of these diseases can be controlled by beginning with clean seed and maintaining a thorough sanitation procedure in handling and cutting operations (details discussed in Chapter 7). It is important to realize that there is a zero tolerance for ring rot in all seed certification programs in the U.S. due to the potential losses from it. Although less severe, blackleg infection can also cause significant losses and must be controlled. Other diseases such as late blight, early blight, and silver scurf must also be prevented. Viruses are controlled by starting with virus-free seed via a meristem or tissue culture program, and by roguing and controlling insects. Seed must be kept as free as possible of spindle tuber, PLRV, PVY, and PVX in order to optimize productivity.

ROGUING

Roguing is used to remove sources of contamination and further increase seed quality; persons walk or ride through fields to identify and remove plants that are suspected of having a disease. Some growers use contracted roguing crews that work large seed areas while others rogue with their own crews. Horses or gas-powered vehicles are sometimes utilized for transporting crew members. Roguing is not viewed favorably by some seed growers because the traffic associated with this process can spread some diseases. Overall, however, this practice has been found to be a valuable aid in producing high quality seed.

SEED PROCUREMENT

The sale of potato seed historically has been a verbal agreement between two parties. In today's business society, this practice is becoming less acceptable. A buyer should provide a contract or utilize the Universal Seed Potato Contract (see Appendix) when closing a seed procurement agreement. These documents provide a written record of most of the salient features of a potato seed purchase agreement.

Transportation of seed to the commercial potato growing regions is by train or truck. The key considerations in transporting the seed are: (1) the use of thoroughly cleaned and disinfected vehicles; (2) maintenance of proper transit temperature; (3) ability of seed to be handled with a minimum amount of damage; and (4) provision for appropriate delivery schedule. The purchaser and seller must agree on the points of seed handling and areas of responsibility for delivery of the seed as a part of the purchase agreement. Each party needs to understand the terms of seed quality and warranty.

Chapter 4

Anatomy and Morphology: Growth and Development

ANATOMY AND MORPHOLOGY

Potato crops are normally planted using small whole tubers or portions of the tuber. These whole or partial tubers are referred to as seed, although they do not resemble true botanical seed. Many cultivars produce significant amounts of true botanical seed in small fruit that look similar to small green tomatoes. The true seed found in the fruit (previously shown in Figure 4) is generally yellow or yellow-brown and is flat, kidney- or oval-shaped, and about 1-1.5 mm long. It is covered with hairs and according to Soueges (1907), the mature seed coat consists of three layers: (1) an internal region consisting of a single layer of epidermal cells; (2) an intermediate zone which is divided into an external and internal region; and (3) an internal or digestive zone which forms the innermost layer of cells covering the endosperm.

The embryo (embryonic plant) is U-shaped and surrounded by endosperm (Hayward, 1938). Upon germination, the radicle (embryonic root) grows rapidly to form a tap root which produces numerous secondary roots. The hypocotyl (stem below the cotyledons, embryonic or seed leaves) elongates, pushing the seed above the soil surface ,and then following expansion of the cotyledons, the epicotyl (stem above the cotyledons) develops. The first leaves to develop are somewhat oval and hairy. The mature leaves may have a series of leaflets along the leaf stalk with opposite arrangement (Figure 8).

It is important to have a basic understanding of the leaf anatomy in order to correctly sample plant parts for nutrient analysis and other diagnostic needs. The petiole is the stem to which the leaflets are attached. The petiolule is the stem which attaches each leaf to the petiole. It is the petiole that is utilized for nutrient analysis.

FIGURE 8. The leaf arrangement on a potato petiole.

Potato leaf

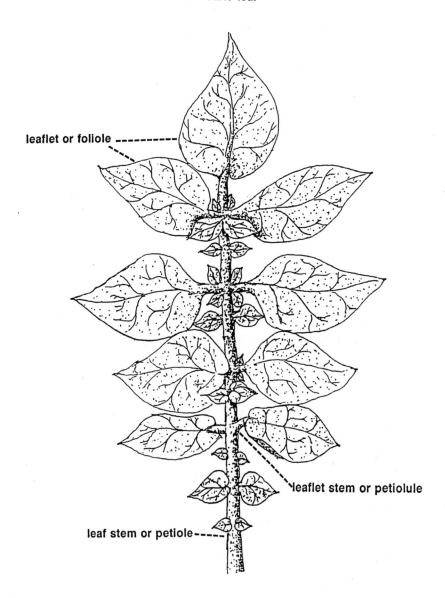

The mature leaf blade contains one layer of cells called palisade cells, and three to five layers of a spongy tissue called mesophyl (Hayward, 1938). The mature stem is triangular or quadrangular in cross section. This shape results from the development of three large vascular bundles and the wing-like projections of the leaf which extend down the stem from each node. The presence, shape, and size of the wing-like projections are characteristic of each cultivar and are utilized in its identification. Between each pair of major bundles there are three smaller ones, and a continuous cylinder of vascular (water-conducting) tissue which is formed by the development of an interfascicular cambium (core of dense meristematic tissue).

The vascular tissue from each leaf consists of five bundles; at the base of the semicircular petiole there are three large centrally located bundles and two smaller ones which lie at its outer edges (Hayward, 1938). The lateral vascular branches in the stem ascend without branching for one or two internodes, while the smaller bundles ascend three internodes before entering the petiole. The vascular tissue is contiguous through the stolon and into the tuber, where it maintains the same arrangement as in the aerial stem. It is difficult to determine which leaves are directly connected to any individual tuber because of this arrangement. However, it is noteworthy that some leaves provide photosynthate disproportionately to specific tubers because of these anatomical constraints. Flow of metabolites in the vascular tissue through the stolon may be at rates of 50 cm/hr (19.7 inches/1 hr) (Crafts and Crisp, 1971).

When whole or parts of seed tubers are planted, all subsequent roots which appear from the stem above the point of insertion in the seed tuber (seed piece) or at points just above the nodes (Figure 9) are termed adventitious roots. The depths to which these roots penetrate depends on the soil type, soil moisture distribution, and cultivar characteristics.

A potato as it is used in commerce is botanically a tuber. The tuber (Figure 10) is an enlarged stolon, which is an underground stem. Both structures maintain characteristics of the above ground stem, such as lenticels (pores), internodes, nodes (commonly referred to as "eyes") and scale leaves. Each node contains several (three to five) auxiliary buds which can form sprouts when conditions for growth are favorable. The tuber consists of four primary zones of tissue. The

FIGURE 9. A typical potato plant during early tuber initiation.

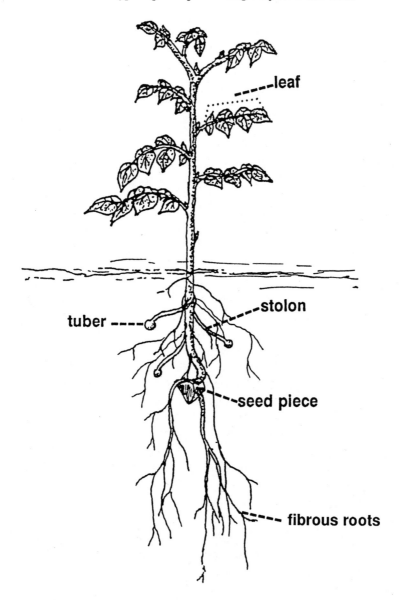

FIGURE 10. The anatomy of a potato tuber.

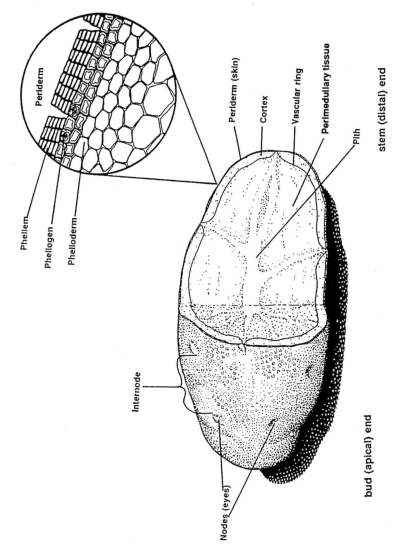

Periderm

Phellem

Phellogen

Phelloderm

Periderm (skin)

Cortex

Vascular ring

Perimedullary tissue

Pith

stem (distal) end

Internode

Nodes (eyes)

bud (apical) end

periderm, or skin, is the first primary zone and is the outermost covering of the tuber. It consists of the phellem (rectangular cells arranged in radial rows), the phellogen (meristematic tissue giving rise to the phellem), and the phelloderm which is beneath the phellogen. The cortex is the second primary zone and lies between the periderm and the vascular tissue. The cortex tissue contains the highest concentration of starch in the tuber. The third is the perimedullary zone, which lies between the vascular tissue and the pith and comprises the largest part of the storage tissue. The fourth zone is the pith, which is the center of the tuber tissue and may be angular with rays extending to each node. The pith is not as high in starch content as the other zones, and thus, cultivars which have tubers with large pith areas are usually less desirable.

The potato flower may be white, yellow, purple, blue, or variegated and is borne on the end of a stem in a cluster. The stamens (male flower parts containing pollen) form an outer cone in the center from which the style (part of the female organs) may protrude (previously shown in Figure 3). The pistil consists of two undiverged carpels which form a two-loculed ovary. The ovary develops into a green, brownish green, or green fruit and may have purple or white streaks. Each fruit may contain 200 to 300 seeds (previously shown in Figure 4).

GROWTH AND DEVELOPMENT

Dormancy and Sprouting

The dormancy and sprouting processes of potato tubers need to be understood in order to manage production practices such as seed storage, planting, and sprout inhibition for marketing or long-term storage of fresh or processing potatoes.

Potato seed producers may influence the sprouting behavior of the potato seed through a number of cultural practices over which they have control. These factors include fertility, water management, planting dates, and storage conditions. Balancing these factors is important in producing good quality seed potatoes.

The potato tuber is vegetative tissue and therefore distinctly different from seeds, which may have several types of rest or dormancy. A freshly harvested potato tuber will usually not sprout when placed in a favorable environment. This condition is referred to as "rest" by Stuart and Milstead, and Wright and Peacock (Hemberg, 1985). Dormancy includes the period following harvest when factors within the tuber prevent sprouting (rest), and the period following rest when factors outside the tuber, such as low temperature, prevent the tuber from sprouting. Although there are wide differences in rest of potato cultivars, no relationship seems to exist between "late" and "early" cultivars and the length of the rest period (Bremner and Taha, 1966). Some *Solanum* species such as *S. phureja* do not have a rest period, whereas some commercial cultivars such as Nooksack may have a very long rest period.

It has been reported that potatoes harvested when immature tend to have longer rest periods than those harvested when mature (Emilsson, 1949). Freshly harvested tubers moved to a favorable environment for sprouting tend to develop sprouts only on the apical end, which become dominant over the other nodes. Potatoes grown in cold, wet conditions tend to have a longer rest period than those grown in a hot dry season, due to a delay in tuber maturity (Borah and Milthorpe, 1962). The longer potato tubers are held in storage, the more rapidly they will sprout when placed in a temperature favorable for growth, and the number of nodes which grow when sprouting does occur will be greater (Burt, 1964). Higher storage temperatures, which speed maturation, decrease the length of the dormancy period.

The implications and interpretation of these results are not straightforward; however, some generalizations may be presented. The rest period is influenced by environmental conditions, either during the growth of the crop or during storage. Any conditions which hasten maturity, such as nutrient and moisture stress or increased temperatures, will reduce the rest period. For example, when the cultivar Nooksack, which has a long rest period, is grown with low nitrogen fertility, it is forced to initiate tubers early followed by earlier than normal tuber bulking, and the length of the rest period is significantly reduced (Dean, unpublished results).

Under conditions of restricted ventilation during storage, it has been shown that 2 to 4 percent carbon dioxide (CO_2) stimulates sprouting, but concentrations above 15 percent inhibit sprouting.

Therefore, it is important to maintain good ventilation in storage so that the CO_2 level stays below 2 percent. Other volatiles given off by potatoes have been shown to inhibit sprouting when they are allowed to accumulate (Bremner and Taha, 1966). Future research will provide a better understanding of what these volatiles are and how they are produced and may allow them to be utilized to control dormancy more precisely.

It has been suggested that plant hormones are involved in the rest period of potatoes. Hemberg (1970) suggested there was a direct correlation between the rest period and the occurrence of specific growth-inhibiting substances in the potato. The inhibitor was found in the peel of potatoes during rest but not in extracts from growing potatoes. The inhibitor was designated the B inhibitor and was the same as the inhibitor B-complex discovered by Bennett-Clark and Kefford (1953) which was later identified as abscisic acid (ABA). El-Antably, Wareing, and Hillman (1967) showed that ABA could be used to inhibit bud growth of dormant potatoes. ABA was later shown to inhibit nucleic acid synthesis, which is probably its major contribution in preventing sprouting (Shik and Rappaport, 1970).

Growth promoters have also been implicated in controlling the rest period in potatoes (Rappaport and Wolf, 1969). Brian, Hamming, and Radley (1955) showed that gibberellic acid (GA) promoted sprouting of potato tubers and Rappaport, Lippert, and Timm (1957) induced sprouting of developing tubers with GA application. Gibberellic acid application to excised potato buds could overcome the inhibition imposed by the inhibitor B-complex (ABA) (Blumenthal-Goldschmidt and Rappaport, 1965) and it was also shown that gibberellic acid application reduced the inhibitor B-complex content of potato tubers (Boo, 1961). The amount of gibberellin in the tubers was shown to increase at the termination of rest (Rappaport and Smith, 1962; Smith and Rappaport, 1961). Wounding of tubers, which was shown to lead to sprouting, was also shown to induce gibberellin synthesis. Since gibberellin treatment was shown to accelerate RNA and DNA synthesis before the onset of cell division or elongation and ABA was shown to inhibit nucleic acid synthesis, it appears that the ratio of these two growth regulators in the cell is the most probable control mechanism of potato rest.

The sprout inhibitor, maleic hydrazide, has been used for many years in potato production. Its mode of action is inhibition of cell division and therefore prevention of sprout growth. CIPC (isopropyl [3 chlorophenyl] carbamate) is also used to prevent sprout growth. Its mode of action is prevention of mitotic cell division.

It seems reasonable that if the process of potato rest is controlled by the ratio of gibberellin to abscisic acid, GA inhibitors or ABA promoters may be used to extend the rest period instead of inhibiting cell division generally. It may also be possible to take advantage of conditions which enhance ABA accumulation in the field in order to enhance dormancy when it is desired. The ABA level increases in many plant systems during stress for water or nitrogen.

Foliage Growth

Once dormancy is broken, stem growth is primarily controlled by temperature. Cultivars such as White Rose and Russet Burbank require temperatures above 10°C (50°F) for any significant growth to occur; the optimum temperature for growth has been found to be between 21°C (70°F) and 24°C (75°F) (Yamaguchi, Timm, and Spurr, 1964). Emergence of potato plants has been shown to be greatly affected by soil temperature. Yamaguchi, Timm, and Spurr (1964) reported that 50 percent emergence of the White Rose cultivar took 8 days at 21°C (70°F), 12 days at 27°-29°C (80°-84°F), 14 days at 16°-18°C (61°-64°F), and approximately 29 days at 10°-13°C (50°-55°F). The optimum temperature, for emergence of the Russet Burbank cultivar is approximately 23°C (72°F) and is very slow below 12°C (54°F) (Dean, 1980). These data should be considered when evaluating the dates for planting. If soil temperatures are below 10°C (50°F), growth will be very slow, if at all, and the seed piece may be subject to pathogen attacks.

At the time of plant emergence, fibrous roots have already been produced at the underground nodes of the stem. Stolon production may also have begun by this time. Although stolons may continue to be produced throughout the plant's life, the major portion of the stolons are produced during the early growing period.

During the time of emergence and initial growth, the plant relies heavily on food reserves in the seed piece. This continues until the plant establishes approximately 200-400 cm^2 (31-62 in^2) of leaf surface area, at which time photosynthesis by the new leaves supplies a greater portion of the carbohydrate needs. Although the seed is providing nutrients and carbohydrates to the foliage, carbohydrates may also be transported to the seed piece from the foliage (Moorby, 1968). This indicates that a dynamic relationship exists between the seed piece and the new plant. Maintaining seed piece integrity by controlling disease with fungicides, suberizing the seed, or proper seed bed preparation is therefore of utmost importance.

Photosynthesis is the process by which carbon dioxide (CO_2) from the air is utilized by chlorophyll in the leaves to make carbohydrates (sugars). This process is probably the most important metabolic event on earth and is certainly the most important process to understand in order to maximize potato production. The most plentiful components in potatoes other than water are the carbohydrates. The carbohydrates consist of the simple sugars, glucose and fructose, the disaccharide sucrose, and the complex carbohydrates amylose and amylopectin that make up starch.

The counterpart to photosynthesis is respiration. Respiration breaks down carbohydrates (sugars) to produce CO_2 and H_2O. It is not the absolute rate of photosynthesis that is important, but rather the relationship between photosynthesis and respiration, termed the net photosynthetic rate. The ability to enhance net photosynthesis to its maximum rate results in the highest productivity.

There are many factors which influence photosynthetic rate and overall production as a result of photosynthesis. A few of the important factors are included here in order to clarify how the process may be understood relative to management decisions.

Selection of a cultivar that has a high photosynthetic rate will potentially result in higher yield if all other factors are equal (Dwelle, 1985). Although some breeding lines or cultivars have shown higher photosynthetic rates, other factors such as tuber quality have prevented them from being accepted and utilized commercially. Climate probably is the most significant factor governing overall productivity.

Maximum net photosynthetic rates occur at temperatures between 20°C (68°F) and 25°C (77°F) (Dwelle, Kleinkopf, and Pavek, 1981; Ku, Edwards, and Tanner, 1977; Moorby, 1968), whereas respiration rates continue to increase well beyond this point. Climates which are best suited to effect maximum net photosynthesis are those which (1) maximize the number of daylight hours and have high light intensity with temperatures between 20°C (68°F) and 25°C (77°F), and (2) minimize the number of hours over 25°C (77°F) and have low night temperatures. The regions which have achieved the highest potato production per unit area thus far are those with moderate day temperature, cool night temperature, and long day length such as found in the Northwest U.S.

The final yield of potatoes is related to the leaf area index (LAI), the leaf area duration (LAD), the rate at which the leaves are photosynthesizing, and how much of the photosynthate produced by the leaves is partitioned into the tubers. Leaf area index (LAI) is the amount of leaf surface per unit area of growing surface. An LAI of "1" would have one acre of leaves over one acre of land. Leaf area duration (LAD) is the length of time photosynthetically active foliage remains on the plant, or simply how long the canopy lives. The LAI and LAD are determined by several factors such as cultivar, fertility, pests, soil moisture, and planting date (these factors and their general effects have been discussed throughout this book). Partitioning (dividing) of photosynthates into various plant parts such as foliage, roots, and tubers can be affected significantly by management decisions and therefore should be thoroughly understood by producers. It is the ability to manipulate partitioning through genetics, fertilizer, and other inputs that has helped achieve high crop yields in many crops including potatoes. Each of these factors–LAI, LAD, and partitioning–are part of the overall photosynthetic system of the whole plant. The control of these factors and the maximum utilization of the climate should be the goals of a potato production system.

Stolon Formation

Stolons are underground stems that develop first at the basal nodes (those closest to the seed piece) of the stem and then develop progressively upward. Any lateral bud can potentially develop either as a leafy shoot or as a stolon depending upon a complex set of factors. The developing stolons can be converted to leafy shoots by exposure to light. This is easily observed in production when the hills in which seed is planted are too narrow and stolons become exposed to light, which causes them to turn green and form above-ground shoots. High temperature during growth may also cause stolons to migrate to the surface and become shoots. Hormones within the plant affect the development of stolons. Gibberellin, in the presence of naturally-occurring or applied auxin (indole acetic acid), stimulates stolon development. A high ratio of another plant hormone, cytokinin to gibberellin favors leafy shoot development whereas a low ratio favors stolon formation (Cutter, 1978). Normal stolon development precedes tuber formation.

Tuberization

Formation of tubers, called tuberization, and their subsequent growth is the most important phase of potato growth and should be thoroughly understood to maximize productivity. There is little satisfaction from producing a beautiful lush green potato field that does not produce a large enough tuber crop to maintain economic viability. To achieve maximum returns, producers must be able to manage crop growth by optimizing the partitioning of photosynthates into the tubers and by making optimum use of climatic and cultural inputs. Tuberization and subsequent growth can be altered by several producer-controlled factors, and therefore returns can be further optimized by proper management decisions.

Tubers form first on the stolons which develop closest to the seed piece (Cutter, 1978). These first-formed tubers may become dominant over the next tuber formed but may not necessarily be the largest tuber on the plant at harvest (Struik et al., 1988). The tuber first appears as a swelling just behind the apex, or growing point, of the stolon. This early development is the result of cell division and cell enlargement and is accompanied by deposition of starch and protein. Tuber formation is controlled by a large number of factors.

Tuberization of potato plants is promoted by relatively low night temperatures (Slater, 1968), short days (<12 hrs) (Gregory, 1956), or stress conditions such as low or interrupted fertilizer supply (Garner and Allard, 1923). High temperatures delay and may even prevent tuberization, which is a major reason why potato production is concentrated in the relatively cool growing areas of North America. The effect of the length of day, or photoperiod, on tuberization of current cultivars is quantitative; short days (10-14 hours) result in tuber production three to four weeks earlier than if grown under long days (>14 hours). Understanding the basis of the responses of potatoes to the climate and to controllable inputs enables a manager to promote early tuber growth for earlier marketing, or delay tuberization until more favorable weather is available to maximize yields.

The terminal portion of the shoot is considered to be the part of the plant which perceives the day length and signals the plant to produce tubers. Gregory (1956) concluded that tuberization results from a stimulus formed or activated by specific conditions of temperature and photoperiod. The stimulus, as yet unidentified, is produced in the actively growing points of the shoot when short days and low temperatures occur. The stimulus moves downward where it induces tubers to form (Kumar and Wareing, 1973). The effect of induction appears to persist even if the inducing conditions are no longer present. The stimulus will cross a graft union but does not move laterally within the plant very quickly (Cutter, 1978).

Application of growth regulators to whole plants and modification of hormones in the culture media used for growing stolons without the influence of the whole plant has shown that hormones play an integral part in tuberization. It has been shown that stolon elongation ceases and tuber initiation occurs when gibberellin levels within the tuber are low (Cutter, 1978). Application of CCC (2-chlorethyl-trimethyl ammonium chloride), which is a gibberellin synthesis inhibitor, to the roots of potato plants stimulates earlier tuberization (Gifford and Moorby, 1967; Gunasena and Harris, 1969). The promotion of tuberization by CCC can be reversed by the application of gibberellin (Tizio, 1969). When gibberellin is applied to potato plants, it causes a delay in tuber formation; if tubers are already developing, it causes secondary growth to occur, presumably by reverting the tuber growth to stolon growth. It has also been shown that radioactively labeled carbon dioxide $^{14}CO_2$ made into sugars by leaves is transported in different quantities to the other foliage tissues or to the tubers depending on the stage of growth when the $^{14}CO_2$ is applied

(Oparka, Marshall, and Mackerron, 1986). If gibberellin was applied to the plants before tuberization, little change in distribution occurred. However, when gibberellin was applied after tuberization occurred, the amount of $^{14}CO_2$ labeled assimilates (sugars and starch) in the tubers was reduced and tuber growth retarded.

Stolons cultured *in vitro* (off the mother plant) were induced to form tubers by kinetin (a synthetic cytokinin) addition to the media, and accompanying the swelling of tubers there was a detectable accumulation of starch, similar to what happens during natural tuberization (Palmer and Smith, 1969). The primary protein found in potatoes (patatin) also accumulates very early in this process and may be an indicator of tuberization (Edilson, Lister, and Park, 1983). The accumulation of patatin may be an effect of tuberization as has been shown with starch, and not a cause of tuber formation. Cultured stolons develop tubers in the presence of kinetin only if temperatures are moderate and sucrose is present. Low temperature treatment or sucrose added to the media without kinetin does not result in tuber formation.

Abscisic acid (ABA) treatment of *S. tuberosum* subsp. *andigena* increased its tuber formation, and treatment of *S. tuberosum* resulted in greater tuber weight (Cutter, 1978). The removal of nitrogen from nutrient cultured potatoes resulted in an increase in ABA content of stolons (Krauss, 1978). When nitrogen was removed under controlled conditions, the interruption of the nitrogen fertilizer supply caused tuber initiation within two days.

Ethylene gas added to the growing environment of cultured stolons, causes swelling of the subapical region, but this is not accompanied by starch accumulation (Kumar and Wareing, 1972). Ethylene has also been shown to inhibit kinetin- or CO_2-induced tuber formation (Mingo-Castel, Smith, and Kumamoto, 1976). Ethylene may play a role in tuber formation by inhibiting stolon growth and maintaining its diageotropic (perpendicular to gravity) growth habit.

Long-term exposure of young plants or stolons to high CO_2 concentrations promotes tuberization (Arteca, Poovaiah, and Smith, 1979; Mingo-Castel, Negm, and Smith, 1974; Patterson, 1975). What role CO_2 plays in normal tuberization is not clear at present.

To summarize thus far, the lesson to be learned from studies on tuberization to this point indicates that the date of tuberization can be manipulated somewhat by cultivar selection (early vs. late cultivars). However, most cultivars currently used in the U.S. (North America) are relatively neutral in regards to day length, so this factor is not a major one in present day production. Controlling or influencing the internal plant hormone levels can be a point of production control by the producer. Although it may not be important to growers to understand the mechanism involved in hormone production, it may be necessary to understand the role of at least two hormone classes, gibberellins and abscisic acid, and what can be done to influence them.

Tuber Growth

The rate of growth of individual tubers on a plant varies, and tubers furthest from the foliage usually obtain the largest size (Winkler, 1971). Competition between tubers for available carbohydrates is strong, as evidenced by the difference in dry matter from tuber to tuber on the same plant (Kunkel, 1966). By using tracer techniques, Moorby (1970) showed that the amount of carbon entering the tubers from the leaves was correlated to the size of the tuber at harvest. There is little doubt that tubers will not be formed on plants which are not transporting carbohydrates to the stolons. However, it is well documented that vine growth is reduced or inhibited when tubers are initiated, indicating that the tubers may be directing the partitioning of carbohydrates (Burt, 1964; Frier, 1977; Nosberger and Humphries, 1965).

The composition of tubers changes during growth. At first, growth and the accumulation of starch is slow, which is followed by a rapid bulking phase when carbohydrates are accumulated rapidly in tubers (Figure 11). Sucrose is relatively high during the rapid growth phase, then decreases until it reaches 1 to 3 percent of total tuber weight. The changes in reducing sugars (glucose and fructose) follow much the same pattern as sucrose. These sugars may increase toward the harvesting period, during moisture stress, or after harvest if stored below 5°C.

The growth of tubers is affected by environment, water availability, and level of soil fertility. The optimum temperature for total vine weight gain is 27°C (80°F) (McCollum, 1978), but this temperature

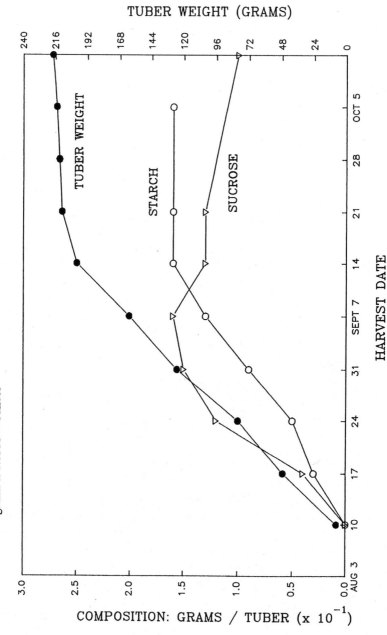

FIGURE 11. Changes in the amounts of sucrose and starch during tuber growth.
*grams x 0.0353 = ounces

66

is too high for optimum tuber growth. The optimal range for tuber growth is 15°-22°C (60-68°F) (Bushnell, 1925; Marinus and Bodlaender, 1975; Yamaguchi, Timm, and Spurr, 1964) and as temperatures rise to 27°C (80°F), tuber dry weight and tuber number decrease. Vine senescence (aging) is also increased at high temperatures. Russeting the formation of a dark, netted skin, of the Russet Burbank cultivar is very poor when soil temperatures fall below 16°C (61°F), although the specific gravity may be high. The starch content of tubers, measured as specific gravity, varies from year to year (Agle and Woodbury, 1968).

Photosynthetic rates usually increase at or following tuberization, which indicates that the tubers partially control photosynthesis. Partial removal of leaves results in increased rates of net photosynthesis from the remaining leaves, and removal of tubers results in decreased net photosynthesis rates (Nosberger and Humphries, 1965).

Moisture stress will reduce plant and tuber growth and subsequent tuber yield from the potato plant by reducing either the fresh and/or dry weight of the tubers. Growth defects resulting from moisture stress may also occur, resulting in reduced economic yield. When moisture stress occurs during tuber bulking, growth constriction of the tuber may result in abnormally-shaped tubers (Robins and Domingo, 1956). Gandar and Tanner (1976) concluded that leaf stresses of –5 bars (1 bar = 1 atm) for three days are sufficient to induce irreversible effects on tuber growth, while stresses of –4 to –5 bars cause cessation of both leaf and tuber growth. Because it is uncommon to find moisture potentials below –5 bars during the day under field conditions, leaf and tuber enlargement occurs primarily at night (Gray, 1973; Kunkel and Gardner, 1965).

Tuber growth (yield) is dependent upon the photosynthetic rate and water uptake. Tuber quality for processing is largely dependent on photosynthetic rate because of its effect on starch content. Potato plants have a broad optimum temperature range for photosynthesis, between 10°C (50°F) and 28°C (82°F) (Dwelle, Kleinkopf, and Pavek, 1981; Ku, Edwards, and Tanner, 1977; Moorby, 1968), above and below which the rates of net photosynthesis decrease rapidly. The optimum light intensity for carbohydrate assimilation is approximately 1200 μ EM^2/sec, which indicates that solar radiation may be a major limiting factor in some growing regions (Dwelle, Kleinkopf, and Pavek, 1981; Ku, Edwards, and Tanner, 1977). When a plant has an LAI of 4.0, i.e., four hectares of leaves for each hectare of land, 80-95 percent of the incident irradiation is absorbed. Each leaf can utilize about 60 percent of full sunlight.

Leaf area duration (LAD) is probably the most important feature of the foliage besides net photosynthetic rate (Baker and Moorby, 1969; Bremner and Radley, 1966). A well-maintained leaf area index (LAI) of about 3.0 to 5.0 provides maximum photosynthetically active surface area and results in rapid tuber bulking. Nitrogen fertilizer application increases LAI and LAD when nitrogen is a growth-limiting factor. For potatoes growing under long sunny days with adequate fertility and moisture, a higher LAI may be more desirable than is appropriate in a short season area with cloudy days because the relatively low amount of solar radiation will not penetrate as many leaf layers. Plants fertilized heavily with nitrogen may become overly vigorous and fail to set tubers, particularly under conditions of low light intensity. If the growing season is short, the tuber bulking phase may be reduced, resulting in poor yields. The average bulking rate of Russet Burbank potatoes growing in Aberdeen, Idaho is 500 kg/ha/day (0.25 tons/acre/day) (Kleinkopf, personal communication) whereas in the Columbia Basin of Washington, bulking rates are approximately 1.1MT/ha/day (0.5 tons/acre/day) (Dean, unpublished). The difference is accounted for by temperature and solar radiation available over the growth period. However, in both cases this bulking generally occurs for a period of about 60 days. Thus, average yields in eastern Idaho of 33 MT/ha 15 T/A (60 days × .25 tons/acre/day) and 66 MT/ha 30 T/A in the Columbia Basin (60 days × 0.5 tons/acre/day) are expected.

The major factor affecting yield and quality is the climate in which the crop is grown. Although climate cannot be controlled, modification of its effects is possible. A manager must determine what the goal of production is. Different practices should be followed for a late crop than for an early crop. Fertility levels, water application, and maintenance of the crop canopy are the most important controllable factors other than cultivar selection. These factors will be discussed individually in the following sections. The reader is encouraged to review the reasons behind the need for production inputs and to develop a system that incorporates inputs because they fit the system, not just as random considerations.

Chapter 5

Cultivation, Fertilization, and Irrigation

INTRODUCTION

Cultivation, preplanting soil preparation, and other tillage practices are extremely variable and usually specific for each producing region, soil type, crop rotation scheme, erosion control problem, energy cost, and other factors. Some general cultivation practices are provided here and which may have broad applications.

Since potato tubers are susceptible to bacterial rots which may be enhanced by anaerobic conditions, any practice which renders the soil anaerobic for any period of time should be avoided. Probably the largest factor causing anaerobiosis is overirrigation. However, proper soil preparation can aid in reducing water ponding or short-term anaerobiosis by allowing good drainage. Deep chiseling or the use of heavy sweeps may loosen hard soil pans which tend to perch water in areas where tubers will form. Each grower should evaluate specific conditions within each field for water perching and other undesirable soil conditions. A hand probe inserted to 45 cm (18 in) will easily reveal these types of soil conditions.

Some potato growers utilize a process of "dragging off" which removes a small part of the top of the hill formed at planting and gives some weed control. This procedure is then followed by a second hilling operation when the plants are 5-7.5 cm (2-3 in) high. Care must be taken to avoid damaging the young sprouts. If the dragging operation breaks the top of the shoot off, it may branch or a new shoot may be produced from the tuber, which will delay emergence. A wounded shoot may also become infected and result in reduced stand.

There has been a significant movement in the West toward reservoir tillage (Figure 12). This technique forms dams and pits (reser-

FIGURE 12. A reservoir tillage device.

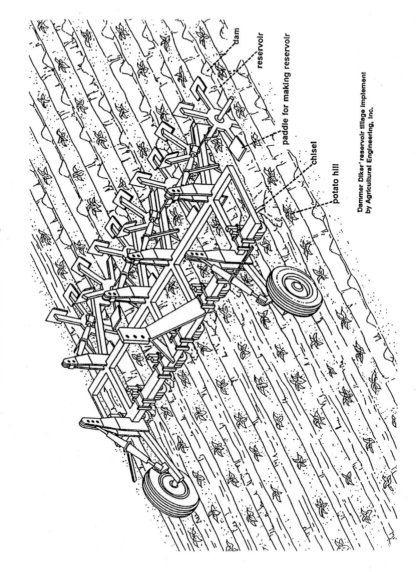

dam

reservoir

paddle for making reservoir

chisel

potato hill

Dammer Diker' reservoir tillage implement
by Agricultural Engineering, Inc.

voirs) continuously in the furrows throughout the field. It establishes reservoirs for water which subsequently infiltrates the soil instead of running down the furrow and out of the field. This practice reduces soil erosion and movement of fertilizer and chemicals away from the intended location. Reservoir tillage results in an increase in water use efficiency, energy savings by reducing pumping costs, decreases pollution of adjacent areas or water systems, and increases efficiency of fertilizer and chemical applications. This operation is generally used at the last cultivation and can be an integral part of the cultivation and insecticide application equipment.

The current timings and current chemicals sections may be used to record current practices and timing for quick reference. A typical ground preparation and cultivation schedule is shown below:

Operation	Timing	Purpose
1. Plowing	As soon as the rotation crop can be harvested. Preplant.	Incorporate crop residue and break up compacted soil areas.
2. Disking	Preplant.	Incorporate fertilizer and prepare seedbed.
3. Chisel and marking out	Preplant.	To loosen soil, apply pre-plant insecticide. Mark rows.
4. Planting	Planting.	Placement of seed, apply-at-planting insecticides and fertilizers and hill formation
5. Drag off	Before emergence.	Weed control.
6. Final cultivation and hilling (use of rolling cultivars, sweeps or shovels with reservoir forming device.)	Prior to row closure.	Weed control, application of insecticides, provide reservoir for water infiltration.
7. Herbicide application	Current timings: 1. _____ 2 _____ 3. _____ 4. _____ 5. _____ 6. _____	Current chemicals: 1. _____ 2. _____ 3. _____

Additional weed control measures will probably be necessary and may consist of sprinkler, ground-, or aerially-applied herbicides. The timing, rates, and appropriate material to use is best obtained from local chemical dealers or Extension Agents. Timing the application of herbicide is critical for effective weed control and is based on many factors including the stage of crop and weed growth, temperature, and anticipated rainfall. Enter your specific timings and chemical applications into the spaces provided in the preceding chart.

NUTRIENT REQUIREMENTS

The potato requires large amounts of potassium and nitrogen for adequate growth. Phosphorus and other minor elements are generally required in smaller amounts. There is some variation in the quantity of nutrients required in various growing regions because of the difference in production potential due to climate, soil types, and disease factors. Generally, the amount of nutrients required is related to the yield of tubers, which is primarily controlled by the environment. Therefore, if the nutrients required for good yields of potatoes in different growing regions are based on a common weight of tubers, a fairly consistent value is obtained. Experimental results from several states were combined in Table 14 to show that the amounts of nutrients removed by the tubers from crops grown in four major production states are similar when expressed as nutrients per metric ton of potatoes.

The figures in Table 14 are for the amount of nutrients removed by the tubers only. The vines will remove 25 percent as much nitrogen, about 20 percent as much phosphorus, and 50 to 125 percent as much potassium from the soil as the tubers. By combining the tuber and vine nutrient uptake data, an average range of nutrient requirements is obtained (Table 15).

The amount of fertilizer which should be applied to the crop is dependent on the supplying power of the soil type, the potential for leaching, denitrification, and cation (positive ion) exchange capacity of the soil, and the growth potential of the crop in the climatic zone. Because of these factors, each potato growing region in the

TABLE 14. Nutrients removed by a potato crop.

	California (Lorenz, 1944, 1947)	Idaho (Painter, 1979)	Maine (Murphy, personal communication)	Washington (Kunkel, 1969)
	(kg/MT^x of tubers)*			
Nitrogen	2.6	4.2	4.2	3.2
Phosphorus	0.7	0.5	0.9	0.7
Potassium	4.7	4.8	5.2	4.3
Calcium	0.1	0.1	0.1	0.1
Magnesium	0.3	0.3	0.4	0.2
Zinc	–	0.003	–	0.002

^{x}MT: metric ton
*To convert to lbs per ton, multiply by 1.9.

U.S. has developed its own set of fertilizer recommendations. Examples are reported in Table 16. An area is left at the bottom of the table for you to insert recommendations that may be different or specific for your area.

FERTILIZER SOURCES

The molecular form in which the fertilizer nutrients such as nitrogen, phosphorus, or potassium are applied can influence crop growth. The growth effect may be due to the rate of uptake of various forms of nitrogen, i.e., little NO_3 is taken up at temperatures below 13°C (55°F) whereas NH^+_4 may be taken up easily (Haynes and Goh, 1978). Recent work with meristem cultures indicates that potatoes prefer the nitrate form of nitrogen over the ammonium form (Davis et al., 1986). The nitrogen forms change rapidly in the soil depending on the presence or absence of certain organisms. It is usually not critical that only one or the other form be used, except if

TABLE 15. Range of nutrients removed in tubers and vines per unit of tubers produced.

	kg/MT	lbs/Ton
Nitrogen	= 4.5–5.9	9.00–11.80
Phosphorus	=0.6–1.1	1.70–2.20
Potassium	=7.1–10.7	14.20–21.40
Calcium	=0.10–0.15	0.20–0.30
Magnesium	=0.25–0.45	0.50–0.90
Zinc	=0.002–0.003	0.004–0.006

the pH value of the soil is marginal and therefore less acidifying forms should be used.

Minor elements such as zinc are required on some new lands in the West. A soil test value of less than 0.8 ppm indicates that zinc may be limiting. An application of 10 lbs/acre would be reasonable if a soil test value less than 0.8 ppm was obtained. Sulfur deficiency can be a problem in some areas and sulfur should be added if needed. Some fertilizer formulations, such as ammonium sulfate, tend to acidify the soil in the root zone, thus indirectly affecting plant growth. Elemental sulfur will reduce the soil pH significantly, and therefore large or continuous additions of sulfur should be avoided.

The use of other minor elements may be recommended in specific areas because of low soil pH, water quality problems, or other reasons. Because the potato crop is of high value, investing in soil analysis and obtaining advice from a consultant on the appropriate fertilization of the crop for a specific soil are easily justified.

Application of potassium in large amounts may reduce the specific gravity of the crop under some conditions and not others (Dunn and Nylund, 1945; Kunkel, Holstad, and Gardner, 1977). Large applications of potassium have also been shown to reduce the susceptibility of tubers to blackspot bruising (Kunkel, Holstad, and Gardner, 1977; Mulder, 1949). These potentially conflicting effects need to be considered in any fertility program. It may be that the KCl form of potassium reduces specific gravity due to the presence

TABLE 16. Recommended fertilizer rates for several potato growing areas.

State	Nitrogen (N) lbs/A	Phosphorus (P_2O_5) lbs/A	Potassium (K_2O) lbs/A
Idaho (Painter et al., 1977)	200	179	224
Washington (Dow, 1974)	320	295	480
Maine ("Fertilizing Potatoes", 1977)	179	280	252
Oregon (Gardner et al. 1985),	340	200	400
North Dakota (Wagner et al., 1977)	67	56	112
Wisconsin (Binning et al., 1980)	202	280	403
Michigan (Vitosh, 1990)			
round whites	170	230	230
Russet Burbank	210		
Florida (Stall and Sherman)			
Irrigated	175	200	200
Nonirrigated	120	160	160
Muck soil	0	160	240
Marl	80	160	240

Current recommendations

for readers' production area: ——— ——— ———

(put your values here)

of the Cl^- ion and therefore, the K_2SO_4 form may be a better choice and at the same time decrease the potential for blackspot.

A worksheet can be utilized to record nutrient needs for individual fields so that fertilizers may be applied based on the best potential returns. Tracking the soil nutrient level over years and maintaining good fertilizer application records can help identify trends toward improper application techniques.

FERTILIZER APPLICATION

The relatively nonmobile elements–potassium, phosphates, and minor nutrients–may be broadcast-applied to the soil and incorporated with soil preparation and follow-up cultivation procedures. Preplant nitrogen should be applied close to or at planting in order to avoid leaching.

Nitrogen application may be done in a number of ways, depending on the equipment available. Single, large applications of nitrogen prior to tuberization may result in delayed tuber set and reduced yields if the growing season is not long enough (see Chapter 4). Split applications can be made at planting and again at row closure, or nitrogen may be applied through sprinklers on a daily or weekly basis. Nitrogen isotope work has indicated that the most efficient uptake of nitrogen is from plant emergence through the early tuber

TABLE 17. The percent ^{15}N recovered in tops plus tubers from N fertilizer enriched with $NH_4{}^{15}N$ applied at different rates.

	% of ^{15}N recovery (tops + tubers) on different sampling dates					
	6/22		7/20		8/17	
	300* N	500 N	300 N	500 N	300 N	500 N
^{15}N application date						
	- - - - - - - - - - - - - - - % - - - - - - - - - - - - - - -					
5/11	34	27	61	44	49	32
6/15	39	34	59	48	54	36
6/29			78	55	60	39
7/13			59	42	48	37
7/27					56	47
8/10					27	15
N rate mean	37	31	64	47	49	34

Roberts and Cheng, 1984
*lbs/acre (Lbs/A × 1.1 = kg/ha

bulking phase (Table 17). The highest percentage of nitrogen was taken up when it was applied before July 13 and when it was applied at the 330 kg/ha (300 pounds) per acre rate.

SAMPLING FOR NUTRITIONAL DIAGNOSIS

Preseason soil tests provide satisfactory information to determine the amount of fertilizer required, when used in conjunction with fertilizer recommendations developed from research in each area.

During the season, soil sampling for nutrients, particularly nitrate and ammonia, will provide a guide to the soil's nutrient supplying power. The availability of these nutrients to the crop is critical for the maintenance of good tuber growth rates. The sampling site should be representative of the major soil type in the field. Samples from each of five to ten locations should be collected from the root zone (first 30 cm (12 in)) and analyzed separately to assess the nutrient variability, as well as the average nutrient content, of the soil. These tests should be conducted weekly on highly leachable soils and biweekly on heavier soils. Obtaining samples from the same site each time allows for better comparison of nutrient availability in the soil from week to week and year to year.

The level of nutrients in the petioles of potato plants has been used to predict the nutrient requirements of the crop. Petioles collected from plants in the same area from which the soil samples are taken are used for the purpose of analytical comparison. The petioles collected should come from the fourth or fifth node down from the plant apex (tip). These will usually be the first fully expanded leaves. Petiole analysis may include all of the nutrients considered to be critical; however, in most cases growers use only the nitrate levels.

The nitrate (NO^-_3) concentration is generally high at the beginning of the season and then drops at the time of tuber growth (Figure 13) (Jones and Painter, 1974; Thomas and Mack, 1938). These tests may be more beneficial in specifying the amounts to use during the next crop than for the existing crop. Responding to current season petiole levels may actually be too late and thus aggravate any nutrient problems. Growers must be aware that peti-

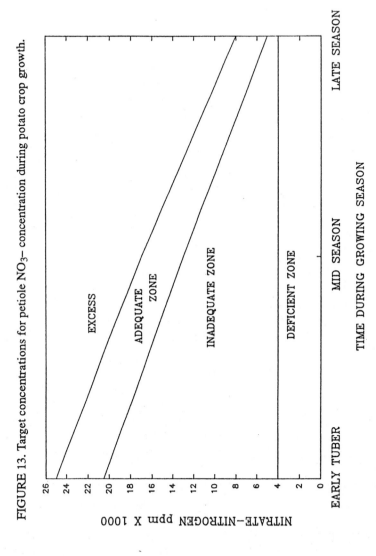

FIGURE 13. Target concentrations for petiole NO₃– concentration during potato crop growth.

FIGURE 13. Target concentrations for petiole NO_3- concentration during potato crop growth.

EXCESS

ADEQUATE ZONE

INADEQUATE ZONE

DEFICIENT ZONE

NITRATE-NITROGEN ppm X 1000

EARLY TUBER   MID SEASON   LATE SEASON

TIME DURING GROWING SEASON

From: Jones, J. and C. G. Painter. 1974.

ole levels change during the day and are affected by climate and cultural practices, not just fertilization practices.

Critical petiole nitrogen ranges or levels have been proposed for potatoes. The levels suggested are generally those above which nitrogen is not considered to be limiting (Jones and Painter, 1974). In Idaho, the recommendation is to keep petiole NO_3^- levels at 18,000-22,000 ppm during early tuber set, 12,000-15,000 ppm during midseason, and 6,000-8,000 ppm late in the season. In Washington, the levels suggested are comparable except late in the season when the range suggested is somewhat higher due to the longer growing season. It is a good idea to use both soil and petiole tests until it is determined which method works best for each producer. Some producers may decide to utilize a combination of the two methods. Where fertilizer cannot be applied after the plants have closed the rows, sampling during the season may still be used to see if nutrients are limiting.

NUTRIENT DEFICIENCIES

The diagnosis of nutrient deficiencies can be very difficult and may be complicated by disease or water stresses. The following guidelines are given to help identify some nutrient deficiency symptoms.

Nitrogen

Nitrogen deficiency symptoms generally occur first on older leaves, which show a general yellowing. The general yellowing proceeds up the plant and will include young leaves under severe nitrogen stress.

Phosphorus

Leaves become dark green and may have a purplish cast when phosphorous is deficient. Foliage growth may be reduced, and leaves and stems may be stiff and oriented upward.

Potassium

The foliage will be darker green in color and may develop bronzing starting from the tips and margins of leaves. The foliage may be crinkled and the veins sunken.

Magnesium

The deficiency is first expressed on lower leaves. They become chlorotic at the tips and margins followed by complete interveinal chlorosis (loss of color).

Zinc

Lower leaves may be chlorotic and young leaves reduced in size when zinc is deficient. The internodes will be shortened and young leaves will have irregular necrotic spots.

Boron

The young leaves and terminal growth show the effects of boron deficiency first. The terminal leaves may be lighter in color and the apex may brown and die.

Calcium

Calcium deficiency is expressed first in the young terminal leaves. They may have a light green band along the margins which will become necrotic, causing them to crinkle or buckle. Axillary buds may show the same symptoms.

Iron

The primary symptom of iron deficiency is loss of color (chlorosis) between the veins. The veins and leaf margins may remain green while the interveinal tissue becomes almost white.

Manganese

Manganese deficiency may appear similar to iron deficiency. Similar chlorosis develops although brown patches develop that may be extensive.

Sulfur

The symptoms of sulfur deficiency are similar to nitrogen except that they appear first on younger leaves. The general yellowing moves from the top downward, without the leaves becoming necrotic.

IRRIGATION

Appropriate application of water during the growing season is crucial to optimize production of the crop. Both the timing and amount of water must be considered in a proper irrigation program. Water should be applied at rates approximately 0.95-1.0 times the evaporation measured from a standardized weather station evaporation pan. A general guide would consist of replacing the water lost from a standardized evaporation pan, and monitoring soil moisture depletion with devices such as a neutron probe, or a tensiometer. Replacing the moisture when it reaches about 75 to 85 percent of the soil water holding capacity will usually meet the crop's needs. An irrigation scheduling program can become a complex model taking many variables into account. Some of the variables which should be considered are soil type, evapotranspiration, crop coefficients, crop growth stage, temperature (present and future), humidity, wind speed, solar radiation, and the ability to irrigate all the crop at one time or on a rotation. There are several irrigation scheduling programs available which can be very effective when combined with visual checks and monitoring. At each growth stage, the crop needs are different, hence, the water use coefficient is different. By multiplying the water use coefficient by the pan evaporation amount, the water need of the crop is obtained. Well-watered plants will have bright green foliage without a bluish cast, and leaves will be firm and not limp. Tubers from plants that are overwatered will have large white lenticels that may become infected by pathogens. Use of these characteristics will help adjust an irrigation program.

Utilization of a basic model for predicting irrigation needs has not replaced the need or desirability of good soil moisture measurements and should therefore be used primarily for long-range (three- to five-day) planning. A schematic chart is provided to illustrate the relationship between evaporation pan data (evapotranspiration) and the calculated water need of the crop (Figure 14).

FIGURE 14. The relationship between crop water needs and evaporation from standard weather station evaporation pan using USDA 'SCHED' for Paterson, WA. 1991. Potatoes planted april 15 and reaching full, effective cover June 19 based on alfalfa as the reference crop.

POTATO EVAPOTRANSPIRATION

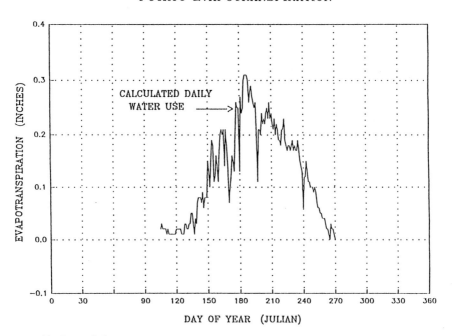

*inches x 2.4 = cm

Chapter 6

Insect Pests

INTRODUCTION

Potato production requires a combination of monitoring and controlling a number of insect pests in order to maintain high productivity. There are several insect species that can cause complete destruction of the crop if they are not controlled. On the other hand, some insect pests may be found in relatively low populations and it may be unnecessary or uneconomical to control them. It is necessary to understand each of the pests and their interaction with the host in order to properly manage the production system.

In general, insect populations respond to increased temperature, light, and other favorable environmental inputs. They may also respond to the availability of a suitable food supply. Providing an environment which favors the crop species and not the pest is a goal of pest management. Attempts to eradicate insects at any level have been shown to be ecologically and economically unsound.

MONITORING PROGRAMS

Insect monitoring can utilize two general techniques: (1) field sampling and (2) insect trapping. The choice of technique will depend on the insect's habits, the number of acres requiring assessment, and the technical skills of the producer. In some instances, a manager of a small acreage may choose to sample a single field, while a manager of several hundred acres may elect to monitor the crop with insect traps if the insect in question can be monitored by traps.

Field sampling must be performed in a way that provides reliable data from which to determine appropriate control measures. First, make a general observation about the field in regards to its topography and proximity to other crops, windbreaks, or native flora. The geography may play a role in insect location by providing shelter from wind, preferred soil moisture regimes, or warmth, as with a southern slope. Second, determine the proximity to other crops or insect hosts that may be important both as sources of the insects themselves or the diseases that they may transmit. Third, establish a monitoring pattern for each field to include features that may be important for locating insects that are potentially destructive. A zigzag sampling pattern is usually preferred and should cover important locations such as the windward side of the field where insects are likely to enter.

The sampling procedure will vary depending on the insects of concern. Sweep nets can be used for large foliage insects but are not effective for aphids. Aphids can be observed on the underside of leaves and in the axils of branches where they may not be dislodged by sweep nets. Large insects such as army worms and loopers may be found by grasping the foliage, shaking it up and down, and observing the insects which fall onto the soil surface. Other specific information on individual insects is provided later in this chapter.

Monitoring fields by trapping is effective for small insects such as aphids, insects that can be lured by pheromones, or for monitoring large areas. Traps are based on the assumption that the insects of concern are moving, which may allow populations of relatively non-mobile insects to develop within the field before they are detected in traps. Pheromone traps, traps that are generally attractive, or preferred host plant traps can help detect insects present in the area.

The appropriate control measure for each insect species must consider (1) the potential loss from not controlling the insect, (2) the number of insects required to cause economic loss, (3) the efficacy of a chosen control measure, and (4) the effect of the control measure on the ecosystem. Current information should always be sought from extension personnel, pesticide consultants, or other certified professionals in the growing area.

IDENTIFICATION AND CONTROL
OF POTATO CROP PESTS

The following descriptions of the major insect pests of potatoes are intended to provide information necessary to develop an overall strategy for pest management. Each insect pest will have unique ecological and physiological characteristics that impact effective control measures. The reader must compile enough information about each pest present in the growing area to develop a strategy for both the immediate need and long-term sustainability. The brief sections following the description of each insect is intended to provide a location to place current recommendations for control of the insect in the reader's locale at the date of reading. Since recommendations may change annually, new information should be included in these sections as it becomes available.

Green Peach Aphid. The green peach aphid, *Myzus persicae*, is a common insect in all potato-growing regions of the United States. This aphid may cause damage by removing fluid from the plant but it is primarily a pest because it transmits potato leafroll virus (PLRV) (discussed in Chapter 7). This makes it one of the most damaging insect pests. The aphid obtains the virus from infected potato plants or other hosts, then the virus passes through the aphid's gut and into its blood. The virus is then transferred to the salivary glands from which the aphid transmits the PLRV via its stylet to the plants. This scenario is important because it means that once the aphid becomes infected with PLRV, it can transmit it throughout its life. This aphid may also transmit other diseases such as the alfalfa mosaic virus.

Green peach aphids (GPAs) overwinter as eggs on alternate hosts such as peach trees (its primary host), plum trees, and other stone fruits. The dark green or shiny black eggs are tolerant to temperatures of $-45\,^{\circ}C$ ($-50\,^{\circ}F$) and are usually laid on or near buds of the above-mentioned fruit trees. Sprays applied in the spring to control other insects are usually effective in controlling the GPA, if used after the eggs have hatched. The wingless females that hatch from the eggs are referred to as stem mothers. They produce a series of generations of live young from which the spring-migrant, winged adults are produced starting with the third generation (Figure 15).

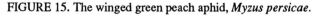

FIGURE 15. The winged green peach aphid, *Myzus persicae.*

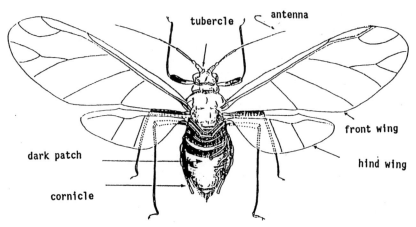

The winged migrants leave the winter host in the spring when temperatures approach 21°C (70°F). They seek out other hosts such as sugarbeet, potatoes, nightshade, mustard, and other species on which to propagate future generations. Nymphs are deposited by spring migrants on suitable hosts and new colonies are produced asexually. New generations are produced in six to 14 days depending on the temperatures. As the colonies become stressed due to crowding, some winged forms are produced which fly to adjacent plants. This cycle is repeated continuously during favorable weather. In the Pacific Northwest, the population of these colonies usually increases rapidly in June, levels off in July, and decreases rapidly in August. In high elevation seed-growing regions, populations may not be significant until August. In the autumn of temperature zone areas, GPAs develop male and female winged stages that mate and deposit eggs on the overwintering host to complete the cycle. In southern growing regions, the aphids may continue in the summer cycle without producing the overwintering egg stage. Producers should be aware of the population dynamics of aphids and other insects in their area and develop the timing of insecticides around the insect's appearances. In the Appendix, a Field Monitoring worksheet is included to graph insect populations in order to determine when insecticides should be applied for insect control.

It is critical to control GPAs both on potatoes and alternate hosts, to plant seed that is free of PLRV, and to rogue out either infected potato plants or alternate hosts near the fields. Preventative sprays may be the only effective means of control in seed areas, whereas monitoring programs can be used effectively in commercial regions. Yellow pans containing water placed on the upwind side of fields on open ground can be used to attract and trap aphids as well as several other insects. The pans should be monitored frequently to keep the water from evaporating and to monitor insect population trends. A small amount of oil can be put on the surface of the water to reduce evaporation.

The green peach aphid can be identified by its characteristic coloration, size, and anatomical features. The aphid body is approximately 2 mm (1/12 in) long, egg-shaped, and usually light green in color, although some wingless forms may be pink. The winged form has a dark brown or black head and thorax with a dark patch on the abdomen. The head has prominent inward-pointing tubercles (Figure 15).

Current control measures

    a.
    b.
    c.

Potato Aphid. This aphid, *Macrosiphum euphorbiae*, can transmit PLRV, and potato viruses Y and A (discussed in Chapter 7) in a nonpersistent manner. This is therefore not as critical a pest as the green peach aphid, but should still be controlled. The winter host for this aphid is the rose, and alternate summer hosts are nightshade, tomato, and ground cherry. The aphid is green or pink and much longer than the GPA. Its tubercles are not convergent and it has long cylindrical cornicles.

Current control measures

    a.
    b.
    c.

Potato Psyllid. This insect, *Paratrioza cockerelli,* occurs throughout the West, but is usually only a significant pest in the central and southern states. The psyllid is a small 2 mm (1/12 in) insect that injects a toxin into the plant while feeding on phloem fluid. The toxin causes the disorder called psyllid yellows, which is identified by the young leaves becoming erect, the basal portions cupping upward, and developing red to purple colorations. These symptoms may be confused with those caused by rhizoctonia infection or purple top virus (both are discussed in Chapter 7). However, additional symptoms on older leaves such as thickening, yellowing, and upward rolling may aid in diagnosis. Aerial tuber formation and the sprouting of below-ground tubers may also occur.

The insects can be monitored with yellow trap pans, sweep nets, or leaf sampling. Normal insecticide sprays may be effective against psyllids, but some systemics are not.

Current control measures

a.
b.
c.

Two-Spotted Spider Mite (*Tetranychus urticae*). Spider mites can be a problem in potato fields in hot growing regions of the West. They are usually not a problem in cool or wet growing areas. The mites are small (< 1 mm (0.04 in)) and a hand lens is required to differentiate species. Mites damage the plants by puncturing the leaves and consuming plant fluids. The leaves are initially observed to have small stippling, which will develop into necrotic blotches that are usually reddish in color. The leaf surface may become glossy and webbing may be found at the leaf or leaflet point of attachment. In severe cases, the plants may become covered with webbing.

The spider mite overwinters as an adult in the soil, on plant debris, or on alternate hosts. The adult female emerges when the soil warms and deposits eggs on the underside of host plant leaves. A single female may lay 300 eggs during her life span. The eggs hatch within three to five days, and develop into adults in seven to nine days.

Mites are spread in the wind, carried about on plant materials, or mechanically transferred. Significant problems seem to occur only during hot, dry periods and usually begin at field margins. Overhead irrigation tends to reduce populations by physically washing mites off plants if the webbing has not become too severe. Some insecticide applications may enhance mite populations by killing predator mites or other beneficial insects. Fields should be monitored for mite infestations, particularly around field edges, and spot applications of appropriate control measures made as needed.

Current control measures

    a.

    b.

    c.

Colorado Potato Beetle. The Colorado potato beetle, *Leptinotarsa decemlineata*, was first discovered in the midwestern part of the United States in the vicinity of the current Nebraska-Iowa border. It has been a persistent pest in potato production since cultivation of the crop began. The degree of infestation varies from region to region and beetle populations from some regions have developed specific pesticide resistance.

The adult Colorado potato beetle is 0.8 cm (1/2 in) long, oval-shaped, and has ten black stripes on its yellow wing covers. The head is red and black. The adult overwinters in the soil, emerges in the spring, and feeds on solanaceous plants. The adults mate and lay eggs on the underside of the leaves of plants such as nightshade, potatoes, tomatoes, or other host plants. The eggs are generally orange or yellow-orange.

The eggs hatch in four to ten days, depending on the temperature, and produce pinkish-red larvae with black spots. The larvae feed extensively for two to three weeks and can defoliate potatoes in a short time. After feeding, the larvae burrow into the soil, pupate, and then emerge as adults in five to ten days, or they may overwinter as adults.

Control of the Colorado potato beetle is best achieved after the eggs hatch and before the larvae mature. Several insecticides are available for control; however, resistances to chemicals have been recorded. Therefore, it is suggested that more than one control measure be incorporated into the management scheme to avoid evolution of further chemical resistance. Biological control measures such as application of pathogenic fungi (*Bacillus thuringensis*) can be used to control this insect. Control by natural enemies is not feasible at this time. Control must be achieved when the larvae are feeding and before they burrow into the soil. If the beetle is in its second generation near the end of the growing season, control may not be economical. A grower needs to determine the value of the potential yield yet to develop versus the cost of the control measure. One alternative approach to treating an entire field is to plant an area of the field with early potatoes without insecticide protection as a trap crop, and then kill the insects when they migrate to it. Insect vacuums and propane flamers have also been used for control of this pest. Appropriate control of alternate hosts, and eliminating cull piles and volunteer potatoes in nearby fields are effective measures in reducing potential sources of the adult phase of this insect.

Current control measures

    a.
    b.
    c.

Cutworms. Cutworms are the larval form of the grey or brown miller moth, commonly seen flying around lights at night. They are nocturnal feeders and usually feed on stems near the soil surface, but also feed on foliage higher on the plant. Symptoms of feeding can be either wilting due to damage of the stem at the soil surface or ragged leaf edges as a result of direct feeding. Cutworms should be monitored by checking in the soil or on the soil surface near injured plants. Loopers and armyworms (discussions follow) can cause similar foliage damage and may be confused with cutworm damage. Cutworms may also feed directly on tubers that are exposed at the soil surface.

Several species of cutworms are found in most potato-growing regions. Common species are the black cutworm, *Agrotis ipsalon;* the variegated cutworm, *Peridroma sacia;* the spotted cutworm, *Amanthes c-nigrum;* and the army cutworm, *Euoxoa ochrogasta.* Although coloration will vary depending on the species, the insects are usually 2.5-5 cm (1-2 in) long, and have a smooth, wet, or greasy appearance.

The moth lays its eggs on stems or on the soil surface; after hatching, the larvae feed actively. One to several generations may occur depending on the length of the season and which species is involved. The cutworm will overwinter as larvae or pupae in the soil.

Biological control with natural populations of parasitic wasps, tachnid flies, predators, and diseases usually keeps cutworm populations in check. Application of chemicals for control of aphids, beetles, or other insects usually provides additional control of this insect.

Current control measures

    a.

    b.

    c.

Loopers. The two common loopers found in potato fields are the cabbage looper (*Trichoplusia ni*) and the alfalfa looper (*Autographa californica*). These insects are usually 3 cm (1 1/4 in) long, green to light green or grey in color with pale stripes on their backs and sides. Because they have two sets of pro legs instead of four sets like other caterpillars, they arch their bodies when they crawl.

The adult looper is a grey moth like the cutworm adult. It lays its eggs in the early part of the season, and may go through several generations before overwintering as a pupa.

Plant injury from the looper is usually found as holes in mature leaves and may be confused with cutworm damage. If loopers are the pest of concern, they can be found on the plant during the daytime, whereas the cutworms feed only at night. Insecticide sprays used for other insects are an effective control for loopers. Natural control is accomplished through bigeyed bug, minute pirate bugs, and other predators which feed on looper eggs. Trichogramma parasites kill looper eggs, and other species such as *Hyposter exigual* and *Capidosoma truncatillum* attack the larvae. Populations may also be reduced by a nuclear polyhedrosis virus.

Current control measures

a.
b.
c.

Yellowstriped Armyworms (*Spodoptera ornithogalli, Spodoptera praefica*). Yellowstriped armyworms can invade potato fields and cause significant damage. The damage is observed as general feeding on the leaves, and complete defoliation may occur. Since these insects usually migrate from adjacent fields, the field margins can be monitored for infestation. Effective control can be achieved with insecticides commonly used for control of other insects such as Colorado potato beetle.

Current control measures

a.
b.
c.

Wireworms. The wireworm is the larvae of the click beetle and is found in most potato growing areas. There are several species which may cause problems in potato production: the sugar beet wireworm, *Limonius californicus;* the Pacific Coast wireworm, *L. canus;* the Great Basin wireworm, *Ctenicera primina;* and the eastern field wireworm, *L. ugonus.* The wireworm larvae are yellowish brown with a dark head and a smooth, tough skin. It is slender, nearly cylindrical, flat on the lower side, and 2.5 cm (1 in) long. The adult is a brown to black slender beetle that gets its name, click beetle, because when turned over on its back, it rights itself with an audible click sound.

After the adults mate, the female burrows into the soil to lay its eggs. The eggs hatch and produce larvae which remain as larvae for three to four years. They may move up and down in the soil in response to soil temperatures. The mature larvae will pupate in the soil in the fall, winter, or spring.

The larval form damages the potato crop by burrowing into seed pieces after planting. The damaged areas may become infected by pathogens and weak plants may result. Larvae may also bore holes into new tubers which increases cullage, particularly in those grown for the fresh market. The hole is usually small, 2-3 mm (1/8 in) in diameter, smooth, and often covered with a brownish periderm.

Wireworms may be present in fields that have produced grain or grass crops in the past even though no crop injury has been seen. The field selected for growing the potato crops can be sampled for wireworms by collecting random samples throughout the field and screening it for the larvae. The field may also be checked by baiting techniques. Pieces of carrot or potato buried 8 cm (3 in) deep may be used to bait wireworms. The bait is left for two or three days in random locations in the field and then checked to see if any damage has occurred. Both soil sampling and baiting techniques will vary in their success based on soil temperature, moisture, and distribution of the insect.

Insecticides are available that are effective for reducing wireworm populations. Once populations have been reduced, reapplication of the insecticide may not be necessary for several potato growing cycles. If grain crops are grown in rotation, additional applications may be necessary.

Current control measures

    a.

    b.

    c.

Leather Jacket or Crane Fly. The larvae of *Tipula dorsimacula* may cause severe damage to potatoes at times. Although feeding is not usually widespread, infested areas may have significant damage.

The leather jacket or crane fly adult is a large, 14 mm (3/4 in) long, slender insect with long-jointed legs and a bright orange abdomen. The adult is apparently attracted to freshly incorporated organic matter (alfalfa) in which to lay its eggs. The grey to greybrown larvae feed in the soil until they mature and form pupae. The pupae move to the soil surface after ten to 21 days and emerge as adults.

The larvae feeding on the tubers leave large 12-19 mm (1/2 to 3/4 in) holes in the surface. The hemispherical, damaged areas are usually clear of debris or frass but may have a rough appearance. Most insecticide sprays are effective for control of this pest.

Current control measures

a.
b.
c.

Potato Leafhopper. Although the intermountain potato leafhopper, *Empoasca filamenta,* is not a serious pest in western states, the *E. fabae* leafhopper found in the eastern and central potato producing regions can cause significant damage.

Potato leafhoppers are small, 12 mm (1/2 in) long, triangular-shaped, pale green insects. Migrant adults are transported on airstreams from southern growing regions to northern growing regions. The females deposit small, slender, white eggs on the leaves of host plants. After six to nine days, the eggs hatch into nymphs that undergo five molting stages (instars) until they become adults. In the North, there may be two generations produced during a season. Populations are favored by warm, damp weather in the presence of an abundant food supply.

These insects cause damage by feeding on the underside of leaves and introducing toxins into the plant. The damage first appears as a small triangular brown spot at the leaf tip. Eventually the entire leaf margin turns brown, causing the leaves to roll upward. This condition is referred to as hopperburn. Insecticides may need to be applied to control this insect if they have not already been applied for other pests. Fields should be monitored and treated before leaves show visual symptoms of hopperburn.

Current control measures

a.
b.
c.

Potato Flea Beetle (*Epitrix cucumeris, E. subcrinita*), **Tuber Flea Beetle** (*E. tuberis)* and **Tobacco Flea Beetle** (*E. hirtipensis*).

The flea beetle adults are small, 1.5 mm (1/16 in), metallic greenish-black to black in color, and hibernate under leaves, grass, or debris on field margins. They chew small holes in leaves that give the plant a shot hole type of appearance. Extensive damage is not usually present, but defoliation can occur. The flea beetle larvae, particularly the tuber flea beetle, can cause significant losses in quality due to burrowing into the tuber cortex. The potato flea beetle's damage usually extends 6 mm (1/4 in) or less into the tuber, whereas the tuber flea beetle's larvae may penetrate up to 12 mm (1/2 in).

The beetles are controlled with pesticide practices used for Colorado potato beetles and aphids. Therefore, specific control measures for the flea beetle are not normally needed.

Current control measures

 a.
 b.
 c.

Garden Symphylan (*Scutigerella immaculata*). Garden symphylans are centipede-like animals. Adults are white, about 1 cm (3/8 in) long, with ten to 12 pairs of legs and a pair of antennae. They live below the soil surface in loose soil and move rapidly away from light when disturbed. Populations appear to be higher in soils with higher organic matter, and infestations have become significant in eastern growing regions.

Symphylans feed on roots and root hairs and can stunt plant growth. They may also feed directly on tubers, creating small entry holes and irregular chambers beneath the tuber surface.

Control may require fumigation or broadcast application of insecticide to the soil prior to planting.

Current control measures

 a.
 b.
 c.

SUMMARY

Economic thresholds have not been established for many insect pests. It should not be concluded that the occurrence of an insect pest requires chemical control. The management of the potato crop will require knowledge of the pest's life cycle, its alternate hosts, the population dynamics of the species, and all potential control measures. Control begins with knowledge of the insect, where it is, how many are present, and what will probably happen if untreated. A thorough sampling protocol for each field should be developed to provide information from which a good decision can be made. Scouting the field on a regular basis can also help the manager assess overall crop condition and intervene where necessary in a timely fashion. Use the Field Monitoring Form in the Appendix to monitor insects and determine if control is necessary.

Consult with county extension agents, Cooperative Extension Specialists, certified professionals, and pesticide company representatives for proper control measures in each area. Record current practices under the Current control measures sections provided throughout this chapter.

Chapter 7

Diseases

INTRODUCTION

Potatoes are confronted with a wide range of pathogens wherever they are grown. The infection of the crop and spread of the disease is influenced to a large degree by the climatic conditions, individual weather episodes, and the susceptibility of the cultivar being grown. Understanding the nature of each disease that may occur on the crop and how it infects the crop is key to knowing how to control its occurrence or spread. Some diseases must be controlled at very low levels of infection, whereas other diseases may be allowed to progress if the plants are at a stage of growth that economic loss is improbable.

The **Current Control Measures** section at the end of each disease description is intended to be for recording current recommendations for control of the disease in the specific growing region of the reader at the date of reading. These recommendations should be reviewed and revised annually as appropriate. From the individual disease control measures, an overall strategy for disease management can be designed.

The diseases are presented in alphabetical order for ease of locating them and not based on their importance.

FUNGI AND BACTERIA

Black Dot (*Colletoctrichum atramentarium*). Black dot is not a serious pathogen in most potato growing areas, although it has recently attracted much attention in the Columbia Basin area of Washington and Oregon. The damage by this organism is still not well documented, but can be severe during hot weather when plants are grown in sandy soils or with marginal nutrition.

The above-ground symptoms include yellowing of the foliage from the top down and wilting. The disease gets its name from the black, dotlike structures called stroma that form on the stems of dying plants. The disease survives as stroma and can become severe when potatoes are grown continuously. A second symptom of black dot is dried pieces of stolon attached to the tuber at harvest. These pieces of stolon result from lesions on the stolon close to the tuber. Sclerotia, drought-resistant hard-walled fungal structure, may develop on the surface of tubers in storage, giving lesions a greyish appearance that can be confused with silver scurf.

The severity of the disease may be reduced through proper crop nutrition, irrigation, and rotation with grain crops. Plant certified seed and avoid infested land.

Current control measures

   a.
   b.
   c.

Blackleg and Bacterial Soft Rot (*Erwinia carotovora*). *Erwinia carotovora* var. *carotovora* (ECC) and *Erwinia carotovora* var. *atroseptica* (ECA) produce diseases called blackleg and bacterial soft rot. Blackleg is produced predominately by ECA whereas soft rot is more frequently caused by ECC. The primary source of inoculum (disease propagules) for diseases is the vegetative seed. The organism has also been found in water supplies, in air samples, and can be introduced by movement of equipment or field workers. While this organism may survive for long periods of time in the soil on plant debris, under most conditions it will not survive.

Erwinia species are favored by cool soil temperatures of 18°-19°C (64°-66°F) and relatively high soil moisture. However, blackleg symptoms may be more prevalent at high soil temperatures 30°-35°C (86°-95°F).

Blackleg is a descriptive term for infection by ECA. Infection is manifested by the appearance of a black, usually slick, appearance of the stem between the seed piece and the soil surface. The infection may continue up the stem, but the plant usually succumbs and falls over before much advancement beyond the soil surface occurs. Other infections may occur at branching points, leaf axils, or points of injury as a result of inoculation from water sources, insects, or other transmission agents. The potato stem will subsequently turn black and slimy, and eventually break and collapse. The foliage will often be reduced in size, and the leaves will yellow and often curl upwards.

Soft rot describes the symptoms of tuber decay by ECC. The infection begins at a wound point, through the rhizome or a lenticel. The bacterium does not have the ability to hydrolyze the epidermal cuticles to gain entry directly. Decay of the lenticellular area results in sunken tan to brownish areas on the surface of the tuber. Internal tissue will usually be cream to tan in color, and can easily be washed out. The rotting tissue may be odorless but becomes quite foul as secondary organisms colonize the tissue.

Tuber decay may be influenced by irrigation, fertilization, hilling practices, harvest damage, and post-harvest treatment. Losses due to *Erwinia* infection can be reduced by using the following protocol:

1. Purchase seed that has been demonstrated to be free of *Erwinia* infection.
2. Disinfect all seed handling and cutting facilities daily or between seed lots.
3. Clean and disinfect planters between seed lots.
4. Avoid planting seed in, or harvesting potatoes from, areas that accumulate water within a field.
5. Reduce irrigation prior to harvest to set the skin and avoid lenticel expansion.
6. Maintain adequate soil fertility and soil moisture.
7. Remove cull and debris piles from the vicinity of the potato field or storage.
8. Grow the crop with large enough hills to protect tuber ends from sunburn.
9. Tarp all trucks to protect surfaces of harvested tubers from direct sun during transport.
10. Wash tubers with clean, chlorinated water and dry them prior to packaging.

Current control measures

   a.

   b.

   c.

Brown Rot (*Pseudomonas solanecearum*). This organism is a problem primarily in the southeastern part of the United States, and has not been reported west of the Rocky Mountains. It is introduced in infected seed and can spread rapidly by seed cutting and planting operations. It has not been commonly associated with cool soils (<15°C (60°F)) although reports have surfaced that show it can survive at lower temperatures. As in the case with bacterial soft rot, tubers from areas of high field moisture are quite susceptible to infections.

Foliar symptoms are wilting, stunting, and yellowing as in the case of *Erwinia* infection. Wilting may be present without any color change. A cross-sectional view of infected stems will reveal glistening beads of a grey to brown slimy ooze in the infected xylem (vascular tissue). Bacterial exudate may be squeezed or drawn out of the tissue. Tubers may not show any external symptoms; however, greyish-white droplets of bacterial slime may be squeezed from the vascular tissue when cut in half. Buds on the tuber may become greyish-brown and exude sticky substances. Root-knot nematodes (discussed later in this chapter) enhance bacterial wilt infection.

Control is obtained by planting disease-free seed, disinfecting seed cutting, and handling equipment and crop rotation.

Current control measures

    a.

    b.

    c.

Common Scab (*Streptomyces scabies, Actinomyces scabies*). The common scab organism is a filamentous bacteria and a low-grade saprophyte that survives on plant material in the soil and may infect a fairly broad host range. It is common in soil that has had large amounts of manure or animal wastes applied and can infect roots of crops such as beets, radish, rutabaga, turnip, carrot, and parsnip as well as potatoes.

The tuber becomes infected through lenticels, although it has been found that the organism is capable of digesting mature periderm *in vitro*, and therefore can potentially invade tuber tissue directly. No foliar symptoms are present even when severe infection of the tubers occurs. The tubers usually develop shallow brown lesions although they can be deep and pitted, depending on the potato genotype and the variety of pathogen. The lesions can cover the entire surface of a tuber when a susceptible cultivar is planted in heavily infested soils.

Control may be achieved by planting resistant cultivars, clean seed, and adjusting the soil pH to below 7.0. Irrigate to provide an optimal level of moisture during tuberization and the early growth period. The organism is favored by dry soil conditions. Some chemical seed piece treatments may reduce scab incidence.

Current control measures

a.
b.
c.

Early Blight (*Alternarea solani*). Early blight is a fungal disease found in all potato growing regions and is most severe in areas of high humidity and frequent rainfall. The disease organism survives on plant debris, in the soil, on potato tubers, and on other host plants. Spores germinate within 35-45 minutes at 24°-30°C (75°-86°F) when the relative humidity is above 96 percent or after two hours of the leaf surface being wet. Infection occurs through the epidermis of either leaves or immature tubers. Wounds or natural openings such as lenticels may be necessary for infection to enter the tuber, as *A. solani* does not appear to produce a suberin- or cutin-degrading enzyme system.

The disease is considered a problem of senescence and if plants are maintained in a vigorous growing state by proper fertility and water application, early blight is not usually a production problem. Symptoms include circular (target spots) lesions primarily on older leaves. Initially the lesions may look like targets, alternating rings of light and dark necrotic tissue, followed by chlorosis and senescence of the leaf. Circular or irregular lesions also form on the tuber surface. The damage is usually contained within the peri-derm and sometimes the cortical tissue. The infection may spread in storage, but secondary infection does not normally occur.

Field monitoring for *A. solani* should be practiced. Prime areas for this disease to begin in the field are high stress locations where nitrogen may be limited, areas where water application is frequent such as the center tower on a pivot irrigation system, or areas where water stress may occur. Maintaining plant vigor through fertilizer application and avoiding stress periods can reduce the problem.

Control is not usually required in areas of hot, dry weather. Cultivars vary in susceptibility and should be evaluated for suitability in the region in question. Protective fungicides may be applied when conditions warrant it and if their success has been proven. Spore trapping with subsequent scheduling of fungicide application is an effective and economical means of disease control. If early blight is present, allow the plants and soil surface to dry following vine killing and before commencing harvest.

Early blight can be controlled by fungicide sprays. The time to spray can be determined by calculating the accumulative heat units required for early blight infection. This procedure utilizes daily maximum, daily minimum, and average temperatures to approximate the amount of growth of an organism. The heat units are calculated by the following formula:

$$\frac{Tmax + Tmin}{2} - base\ T$$

Tmax = maximum daily temperature
Tmin = minimum daily temperature
base T = base temperature (below which growth does not occur), usually (10°C)50°F

If this procedure is used in the San Luis Valley of Colorado, the first blight spray application is made when the accumulated heat units reach 360 degree days (°C; 650 degree day (°F)); or in northeastern Colorado, 625-639 degree days (°C; 1125-1150 degree days (°F)) with a 50° F base.

Spore traps used to monitor the presence of the organism in the area can also be used to evaluate the potential for the disease and to avoid losses or unnecessary sprays.

Current control measures

a.
b.
c.

Fusarium Wilt and Dry Rot (*Fusarium* species). This pathogen is found in virtually all soils, and species of *Fusarium* are pathogens on most agriculturally important crops. The diversity of this organism is apparent by the number of species that attack potatoes.

Foliar wilt diseases are caused by *Fusarium eumartii, F. oxysporum, F. avenaceum,* and *F. solani.* Tuber dry rots are caused by *F. roseum* and *F. solani.* The organisms differ in optimum temperature for growth, in color, mycelia fungal body, and other characteristics, but cause similar effects on the plants. Specific differences between the species can be obtained from the *Compendium of Potato Diseases* (Hooker, 1981). The fungal organisms are primarily located in the vascular tissue and therefore cause wilting of the plant. Eumartii wilt is the most severe of these diseases and is characterized by the yellowing of the youngest leaves between the veins followed by necrosis and bronzing. Wilting caused by *F. oxysporum* is usually rapid and the yellowing of leaves begins lower in the plant.

Plants infected with *F. avenaceum* may be dwarfed early in the season and produce aerial tubers with associated purpling of the foliage. Wilting may predominate on one side of the plant.

F. solani causes rotting of the roots, the stem pith, and the underground stem. Foliage symptoms include wilting, yellowing, and the appearance of aerial tubers.

Dry tuber rots caused by *Fusarium* species appear first as small brown areas around wounds. It has been reported that *Fusarium* cannot enter the periderm directly, although some forms of *F. solani* can produce enzymes capable of hydrolyzing suberin. Naturally-occurring infections appear to come only through wounds created by harvesting and handling. The decay spreads slowly in the field or in storage to produce tubers that have wrinkled, soft, spongy, and usually dry regions on one end or the other of the tuber. Secondary infection by bacteria may make the rotting area slimy or watery and foul smelling.

The wilt diseases are more prevalent in the midwestern and eastern parts of the United States. Dry rot can be a problem wherever potatoes are grown. Avoid damage at harvest and apply a fungicide if needed at the time tubers are placed into storage. Do not plant *Fusarium*-infected tubers.

Current control measures

    a.

    b.

    c.

Late Blight (*Phytophthora infestans*). The disease, late blight, caused by *Phytophthora infestans*, is the most famous and widely spread of the potato diseases. This fungal disease resulted directly or indirectly in the deaths of one million people and the emigration of one and a half million people during the Irish potato famines between 1846 and 1860. It occurs in every potato-growing region of the world and is a consistent and serious threat in cool, wet climates.

The organism, *Phytophthora infestans*, overwinters or survives on susceptible plant material such as potato tubers in the soil, in storages, near fields, or in cull piles. The sporangia (spore-type structure) germinate and form either a germ tube or zoospores. The germ tube of the sporangia or zoospores attach themselves to the plant surface and penetrate to cause infection. Germination occurs only if the plant surface has been wet or when the relative humidity of the air is above 90 percent and the air temperature is below 25°C (78°F). Following penetration, the fungi produce haustoria (penetrating feeding organs) that invade surrounding tissues. Leaf symptoms vary depending on the cultivar, but are usually easy to discern. Leaves show dark green areas that appear water-soaked and have a surrounding light green halo. When the temperature and moisture conditions are satisfactory, white sporangia may be readily seen, particularly on the undersides of leaves. When viewed from the top, leaves often have a greyish-green circular region, and the same location on the undersides of leaves will have the sporangia. If the disease is not controlled promptly, the foliage will deteriorate rapidly as the disease spreads. The leaves and stem will turn dark green to grey, then to black and die. Complete foliage death can occur within a few days.

Tubers may become infected if they come in contact with spores through soil cracks, harvesting, or other means. If a tuber becomes infected, it will show irregular, slightly depressed brown to purplish areas on the skin. The tuber rot will generally be somewhat dry and brownish in color. Secondary organisms may infect partially rotted tubers and result in extensive losses.

A total disease management scheme should be practiced to control this disease. Fields, storages, and equipment should be kept clean of potato debris. Cull piles should be treated to prevent growth and buried to prevent aerial spread of sporangia. Volunteer potato plants in the field or in adjacent fields should be killed to prevent them from being a source of inoculum. Certified seed from disease-free growing areas should be planted. Fields should be monitored for the occurrence of the pathogen and a program for late blight forecasting used if available.

Appropriate protective fungicides should be applied in areas where late blight is a perennial problem. Since this organism has shown a propensity to develop chemical resistance, use of more than one chemical is recommended. In the dry areas of the West where blight occurs infrequently, remedial sprays may be effective. The most common areas where late blight can be found in center pivot irrigated potato production is within the first section nearest the pivot. Oftentimes this area has a high relative humidity and the foliage remains wet for long periods. Late blight control may be achieved by periodically or routinely turning off the nozzles in this section for short periods (one revolution).

Potato vines should be allowed to thoroughly desiccate or should be removed prior to harvest to avoid tubers coming in contact with infected tissue. Allowing the soil to dry slightly may also reduce potential infection.

Storages should be cleaned and air systems provided before tubers go into storage (discussed in Storage section of Chapter 9). If late blight has been found in the field, it may be useful to dry the tuber surfaces by running the storage air systems without humidification for a short period. The amount of weight loss due to water loss can be significant under these conditions, so a manager must evaluate the potential for weight versus the potential for rot loss.

Current control measures

   a.
   b.
   c.

Leak (Water Rot) (*Pythium ultimum, P. debaryanum,* or *P. aphanidermatum*). Leak is caused by infiltration of the fungus into tubers, through wounds which occur at harvest. The disease is usually more prevalent following high temperatures at harvest (*P. ultimum*) and if the soil moisture is high during a hot harvest period (*P. aphanidermatum*).

Tubers are infected through wounds, and upon storage, the breakdown of internal tissue occurs. If the infected tuber is cut in cross-section, the surface will oxidize rapidly to initially produce a reddish pigment followed by the tissue turning black. This is similar to the symptoms of pink rot (discussion follows). The leak symptoms will appear primarily in the center of the tuber and may leave a ring of sound tissue around the periphery. When pressure is applied to the affected tissue, it bursts open and exudes a watery substance. Tissue infected with pink rot is usually spongy and does not release water so readily.

The organism lives in the soil and can be controlled by avoiding high soil moisture prior to harvest (*P. aphanidermatum*) and maintaining high humidity following harvest (*P. ultimum*).

Current control measures

   a.
   b.
   c.

Pink Rot (*Phytophthora erythroseptica*). Pink rot causes tuber symptoms much like leak (see above), but also causes wilting and leaf abscission of the foliage. Infection usually occurs through the stolons, but may also enter through lenticels or the eyes. The rotted tissue is delineated by a distinct dark line and looks grainy when cut open. The tissue will turn pink and then black within an hour after cutting.

The rot is aggravated by high soil moisture late in the season, especially when hot weather occurs. Unlike leak, pink rot can spread in storage. Maintain sufficient air flow in storage and reduce temperature to near 4°C (40°F) to reduce spread of infection. Since the disease is favored by high soil moisture, provide drainage where necessary and avoid overwatering, especially late in the season.

Current control measures

    a.
    b.
    c.

Powdery Scab (*Spongospora subterranea*). This disease has become more important in western growing regions during the past five to ten years. It was first identified in eastern Canada in 1913 but is now reported in most potato growing regions. The organism belongs to a class of slime molds (*Plasmodiophorales*) which are classified as obligate endoparasites.

The organism survives in the soil as spores which germinate to produce zoospores in the presence of susceptible host plant roots. The zoospores enter the roots or tubers through natural openings or wounds. The root tissue enlarges to form galls (abnormal outgrowths) which are characteristic of the disease. When the galls mature, they change from a whitish color to brown and eventually rupture to release resting spores.

Symptoms of tuber infection are at first seen as small, raised, brownish areas about the size of a pinhead. The areas of infection then expand and may take on a jelly-like appearance, followed by drying and splitting of the epidermis. Characteristic symptoms include a crater-like scab with the epidermis peeled back. This differentiates it from common scab, which usually has rough-edged lesions. The crater will also contain masses of dark spores.

The disease persists in the soil, and resistance to it is not currently available. Rotation with non-host crops is the only currently available control.

Current control measures

a.
b.
c.

Rhizoctonia, or Black Scurf (*Rhizoctonia solani*). This disease is spread worldwide and the tuber symptom has been referred to historically as "the dirt that won't wash off." This description comes from the fungal masses (sclerotia) produced on the tuber surface by *Rhizoctonia solani*.

R. solani is a fungal disease that infects the underground plant parts, but symptoms may be seen on the foliage. The underground portions of stems and stolons of infected plants will have reddish-brown cankers of irregular size and shape. If soil temperatures are cool (< 12°C (54°F)) at planting and during preemergence, infection may be severe enough to cause sprout death. Tuber malformation may occur and the small black sclerotia may be present in large numbers on the tuber surface.

Foliar symptoms usually result from severe infection of below-ground stems. The leaves may show yellowing or reddish coloration particularly in the top section of the plant. Leaves may curl upward and aerial tubers may form in the axils. These symptoms can be induced by several organisms and are general physiological responses to girdling of the stem. Whitish-grey fungal growth may occur on the stem just above the soil line of infected plants, but often is not present. This is usually easy to rub off.

Complete control is not possible at this time; however, the severity can be reduced by planting disease-free tubers in warm soil that is not too wet, or by some chemical seed piece treatments. Rotation crops can be helpful, but large amounts of undecomposed residue may increase disease severity.

Current control measures

a.
b.
c.

Ring Rot (*Clavibacter michiganese* subsp. *sepedonicum*) (formerly *Corynebacterium sepedonicum*). Ring rot, produced by the bacterium *Clavibacter Michiganese,* can be a very destructive pathogen given optimal growing conditions. It is the only disease organism for which a zero tolerance has been implemented in Seed Potato Certification programs.

C. michiganese is a gram-positive bacteria that overwinters in infected tubers, on burlap bags, storage walls, plant debris, equipment, etc. The pathogen is spread in seed-cutting operations and during the planting operation. The inoculum from one infected seed tuber may be spread to many succeeding seed pieces depending on the inoculum level. The optimum soil temperature for disease development is 18°-22°C (64°-72°F). Disease expression varies greatly between cultivars; late-maturing, vigorous cultivars may have considerable tolerance to the disease.

Symptoms include wilting, rolling of leaf margins, and general yellowing of older leaves. A dwarf rosette appearance of new foliar growth is common with the Russet Burbank cultivar. Distinguishing characteristics of tuber infection are the bacterial ooze which can be squeezed from the vascular system, similar to that with brown rot. The coloration of this ooze may be more creamy compared to the greyish color with brown rot. The tuber surface will often crack as a result of internal pressures generated by the bacteria, which is not common with brown rot.

Current control measures

    a.
    b.
    c.

Silver Scurf (*Helminthosporium solani*). Silver scurf occurs in all potato-growing regions. It is generally of secondary or little importance but can cause significant losses. The fungal organism is transmitted primarily on diseased seed, but can carry over in the soil. Infected seed pieces produce conidia, the spore containing organ, which sporulate and produce new mycelia, which then infect new tubers directly through the periderm or lenticels.

The disease is characterized by the shiny, silvery appearance of small, light brown or greyish leathery spots. Because the disease usually originates from infected seed, these silvery areas will appear first on the stem end of the tuber, nearer the infection source. The margins of the lesions may appear sooty due to the production of spores. The fungus is capable of spreading in storage, particularly under high humidity and warm temperature conditions. Tubers may become covered with the sooty appearance from spores during storage, and decay and weight loss will be obvious from the tuber shriveling that occurs.

Methods of control include planting disease-free seed in soil in which the disease has not been present in previous crops. The organism does not have other hosts, so only potato crops need to be checked for the disease for rotation management. Harvest should proceed as soon as the crop is mature, to remove the tubers from conditions that favor the disease. The storage temperature should be lowered as soon as wound healing has occurred, to minimize sporulation and spread of infection. Although reducing the relative humidity to less than 90 percent may reduce sporulation and disease spread, this practice may also increase pressure bruise due to weight loss and cause other long-term storage problems.

Current control measures

   a.
   b.
   c.

Verticillium Wilt (*Verticillium dahliae, Verticillium albo-atrum*). Verticillium wilt is caused by two species of Verticillium: *V. albo-atrum* and *V. dahliae*. Both species may be present in the plant at one time, and the plant may not show symptoms even when significant amounts of the fungus can be isolated from the tissue. The disease organism survives in the soil either as mycelia in the case of *V. albo-atrum*, or as pseudosclerotia with *V. dahliae*. The pseudosclerotia (sclerotia like structure) last for longer periods in the soil (eight to ten years) and therefore, longer rotations are necessary when this species is present.

The source of the fungus is either seed tubers or alternative hosts, of which there are many. Clean seed should be purchased and seed treatment should be considered to reduce the spread of disease. The fungus grows best at relatively low temperatures, with *V. albo-atrum* favored at lower temperatures at 16°-27°C (60°-80°F), compared to

V. dahliae at 22°-27°C (72°-80°F) and neither may germinate at a high temperature of 35°C (95°F). The organism enters the plant primarily through root hairs, wounds, and sprout or leaf surfaces. It penetrates cells and enters the vascular system (xylem) where it causes its primary symptoms.

Internally, Verticillium may cause the xylem to become brown to reddish-brown in color. This symptom is often found at the soil surface areas of the stem and can be found on the stem end of the tuber. Significant discoloration of the vascular tissue may occur. The tissue remains solid, unlike tissue infected by *Erwinia* species or ring rot bacteria. The damage to the xylem tissue results in wilt symptoms from which the disease gets its name. Wilting may include the entire plant early in the season, but usually affects one stem or a portion of one stem. By mid-season, the symptoms appear as one or more upright stems, usually noticeably erect compared to other stems on the same plant. The first leaves to show symptoms are the older leaves followed by a general wilting, yellowing, and death of all the leaves on a stem. The symptoms may occur during the middle of the season and therefore have been referred to as "early die." The early death of plants can be caused by a number of individual diseases or combination of organisms, so care must be taken to thoroughly assess the cause.

Some cultural practices significantly influence the development of this disease. Plants that are stressed for water or nitrogen are much more susceptible to Verticillium infection or symptom expression. It is more likely that stressed plants are more susceptible to expression of the disease because the organism can be readily found in healthy plants; it is also reasonable to assume that water stress in particular will aggravate a vascular disease problem. Cultivar selection can influence the effect of the disease. Some cultivars appear to be more resistant, although reports conflict about the susceptibility of certain varieties, such as Kennebec (compare Hooker 1981 vs. IPM for Potatoes); and, the author's observation with cultivars such as Nooksack has shown that although some cultivars are reported to be resistant, they may be susceptible when grown with low nitrogen to produce an early crop.

The severity of the disease can be reduced by using practices that optimally maintain foliage growth. Adequate water and nitrogen are the two most important factors, but nitrogen rates that are too high can cause delayed harvest and other problems. Choose cultivars that are more resistant to the disease. Fumigate crops to reduce the disease directly, or to reduce damage done by nematodes such as *Pratylenchus penetrans* (see Nematode section in this chapter), which have been shown to increase severity of Verticillium wilt.

Current control measures

a.
b.
c.

Wart (*Synchytrium endobioticum*). The potato wart disease is extremely serious in some parts of the world, and its spread has been limited by legislative quarantines. Although it is not found in most commercial potato growing areas in the United States, fieldworkers should be able to identify it.

The organism germinates in the soil from resting spores to produce zoospores. These zoospores rapidly penetrate the epidermis and cause subsequent proliferation of underlying cells. The proliferation results in the characteristic warty outgrowths or tumorous galls which are characteristic of this disease. Galls may be found on above-ground plant parts as well as on the tubers, but usually not on the roots. Galls are normally green to brown on foliar tissues and white to brown on tubers or stolons, but will turn black due to decay.

The disease is very persistent in the soil and no effective chemical control is available. Resistant cultivars should be planted in areas where this disease is present.

Current control measures

a.
b.
c.

White Mold (*Sclerotinia sclerotiorum*). This disease is present in most potato growing areas and is predominate in cool climates. The organism, *Sclerotinia sclerotiorum*, overwinters in the soil or in plant debris as hard fungal bodies called sclerotia. These usually have a black exterior, are oval to irregularly shaped, variable in size, and are white or black on the inside. Late in the season, the sclerotia can be found in the potato stem near the soil surface where infection has occurred. Tubers can be infected if they are exposed, but this is rare.

The sclerotia germinate under moist conditions and cool temperatures to produce light brown, funnel-shaped or flat fruiting structures called apothecia. Spores are produced by these structures and are dispersed by wind to other plants. The spores will germinate and cause new infections if free moisture is present for at least 48 hours. The leaf symptoms include water-soaked areas, and a white mycelial material may be present.

The disease can be controlled by reducing irrigation frequency, application of systemic fungicides, and by long-term rotations with grain crops.

Current control measures

a.
b.
c.

VIRUSES

Alfalfa Mosaic Virus–Calico. Alfalfa mosaic is easily recognized by the bright yellow symptoms called Calico which appear on the upper portions of the plant following infection. The disease is spread by 16 species of aphids, but predominantly by the pea aphid, *Acyrthosiphon pisum*. Alfalfa plants are the source of the disease. When the alfalfa is cut, aphids move to adjacent crops, including potatoes, and spread disease.

Foliar symptoms are bright yellow coloring of new leaves and some stunting. The stems may contain some necrosis and, if severe, tubers may show dry corky areas or patches.

This disease can be reduced by not planting next to alfalfa fields, especially older alfalfa fields. Seed growers should rogue all infected plants, and commercial growers may want to consider roguing even though the disease will not be further transmitted from the infected plant.

Current control measures

   a.
   b.
   c.

Potato Leafroll Virus. This virus is probably the most wide-spread and serious of the potato viruses. It occurs throughout the world and can cause serious yield and/or quality losses. Potato leaf-roll virus (PLRV) is transmitted by the green peach aphid, *Myzus persicae*, as well as other aphids (see Chapter 6).

The infection is located only in the phloem cells, one of the conductive tissues of the vascular system. These cells become plugged which results in the accumulation of starch in the leaves, giving them a rigid, papery feel and the diagnostic rustle when the plants are shaken. "Current season" PLRV infection results in younger leaves becoming infected first, followed by spread to lower leaves. The leaves turn yellow, roll from the edges, and may have pink or red pigmentation on the leaf margins.

Tubers become infected if foliage infection has occurred before tuber growth stops in the fall. Brown streaks, called net necrosis, become evident in the tuber vascular tissue. Net necrosis is prominent in some cultivars such as Russet Burbank, Norgold Russet, and Green Mountain.

Control of the PLRV or its symptom development is achieved by purchasing clean certified seed, controlling both the aphid population and volunteer potatoes, and not storing potatoes from infected fields.

Current control measures

   a.
   b.
   c.

Potato Spindle Tuber Viroid. This disease is caused by a small RNA (ribonucleic acid) molecule similar to a virus except that it does not have a protein coat. The potato spindle tuber viroid (PSTV) causes young stems and flower pedicels to be longer and more slender than usual. Leaves and leaflets tend to maintain a more upright growth habit and growth is reduced. Leaves may become rugose, have a rough uneven surface, and twisted with severe strains. Tubers show the most prominent symptoms. The tubers may be elongated with pointed ends, deep eyes, and prominent eyebrows. Skin color and russeting may be reduced, and surface cracking parallel to the long axis is frequent. Internal tuber tissue may have necrotic areas.

Current control measures

    a.
    b.
    c.

Potato Virus S. Potato virus S (PVS) is a symptomless virus and is present in many or most seedlots in the United States. The virus does not seem to cause any effect by itself, but may cause yield reduction when combined with PVX.

Potato Virus X. This virus (PVX) may show mild mosaic dwarfing of the plant, rugose leaves, or no symptoms. PVX is transmitted mechanically and carried in the tubers. Fields should be inspected for this virus during periods of lower light intensity such as early morning or cloudy days. Seed production must rely on the testing of plants during the season by ELISA (enzyme-linked immunosorbent assay) or other tests to prevent PVX infection.

Current control measures

    a.
    b.
    c.

Potato Virus Y. This virus (PVY) can result in serious crop losses. The disease is transmitted by a large number of aphids but most significantly by the green peach aphid. Symptoms include mosaic coloration of the leaves and a darkening of the veins on the underside of the leaves. When both PVY and PVX infection occur at the same time, the disease can be severe. Symptoms include rugose mosaic where the leaves are rough in appearance, irregular in color, and reduced in size.

Control of PVY is achieved by planting certified seed, controlling aphids, and eliminating volunteer potatoes and alternate weed hosts.

Current control measures

a.
b.
c.

Potato Virus M. This virus is not of economic significance in North America. It appears to be present in much of the potato seed produced for certification, but does not cause economic loss. The leaves of infected plants may become mottled or crinkled with mosaic coloration or leaf rolling.

Control aphid populations and plant virus-tested seed to minimize the amount of this virus in the field.

Current control measures

a.
b.
c.

Potato Virus A. Potato virus A can cause severe losses of yield. Leaves may have a mild mosaic pattern or develop severe chlorotic mottled areas with alternating yellow and green colorations of irregular shape. Control of PVA is achieved by planting clean seed of resistant cultivars and controlling aphids.

Current control measures

a.
b.
c.

Tobacco Rattle Virus. Tobacco rattle virus (TRV) is also called corky ringspot, spraing, or sprain. Corky ringspot is a good descriptor for some susceptible cultivars. The stubby-root nematode (*Tri-*

chodorus or *Paratrichodorus,* discussed later in this chapter) transmits the virus, and once transmitted, the disease results in the formation arcs or concentric rings of corky, necrotic tissue in infected tubers. The foliage will not usually show symptoms.

Because the disease requires nematodes to spread infection, eliminating them by fumigation, nematicides, or rotation helps control the disease. Planting certified seed and controlling weed hosts also helps to reduce the disease.

Current control measures

   a.
   b.
   c.

MYCOPLASMAS

Aster Yellows, Purple Top Wilt, Haywire, Potato Late-Breaking Virus. Although this disease is known by several names, aster yellows is the most common. It is caused by a mycoplasma which is a simple organism like a bacteria. It is made of a single membrane containing strands of fibers that are presumed to be DNA. It is transmitted by the aster leafhopper, *Macrosteles fascifrons.* Alternate hosts are numerous including vegetable, ornamental, and field crops as well as weeds such as Russian thistle.

Purple top is a good descriptor for this disease because its primary symptom is intense purpling of upward-curling leaves in the top of the plant. Aerial tubers are usually present in stem axils. The tubers may have necrotic tissue scattered throughout the tuber and may be reduced in size. Normally, the mycoplasma does not survive the storage period and therefore plants arising from the infected seed appear normal. If the mycoplasma survives storage, weak sprouts (hair sprouts) are produced.

Control of this disease is only through seed certification procedures and leaf hopper control.

Current control measures

a.
b.
c.

Witches Broom. This mycoplasma-caused disease is not a significant problem but can occur in most potato growing areas. The disease is spread from alternate hosts (but not from other potatoes) by leafhoppers, *Scleroracus* sp. The disease survives in tubers, so when infected tubers are planted as seed, plants with diagnostic symptoms are produced. The leaves are usually simple and rounded and many stems are produced. The tubers are usually very small and therefore the disease eliminates itself.

Control is achieved through seed certification procedures and control of leafhoppers.

Current control measures

a.
b.
c.

NEMATODES

Nematodes are small microscopic roundworms that attack potato roots and tubers. There are several types of nematodes, and the locations in which each type is found are related to its ability to survive and reproduce in each climate. They may cause direct damage to roots and tubers resulting in yield or quality loss, or may provide entry points for other diseases.

Potato Cyst Nematode, Golden Nematode (*Globedera rostochienis, G. pallida*). This nematode gets its name from the hard round cysts (mature females) it produces which encapsulate the eggs. The larvae hatch in warm weather, enter the roots, feed and develop until they enlarge enough to rupture the root tissue. The female remains attached to the root while the males leave the roots and mate with the females. Eggs are produced within the female and the skin (cuticle) of the female darkens and hardens to form a cyst attached to the root. The nematode is best suited for relatively cool climates as indicated by its infection optima of 16°C (60°F).

Since the nematode can survive in soils for 20 years or more as a cyst, eradication is impossible at this time. Quarantines have been established in areas where this nematode has been identified to prevent its spread. Potatoes should not be planted in soil having the cyst nematode, and all equipment should be thoroughly sanitized when moving out of a field which has, or is suspected of having, this nematode and practice long rotations. Always purchase certified seed to avoid this pest.

Root-Knot Nematode (*Meloidogyne chitwoodi, M. hapla, M. incognita*). These nematodes are pests of potatoes and a large number of other crops throughout the world. Because they are able to infect a large assortment of plants, control by rotation is very difficult. The distribution of *Meloidogyne* species is dependent on temperature and can be approximated by knowing the optimum temperatures for activity, reproduction, and infection. Species such as *M. chitwoodi* (Columbia Root Knot) are found in many production areas that are relatively cool such as Washington, Oregon, Idaho, Colorado, Utah, Nevada, and coastal California. *M. hapla* is found additionally in central California while *M. incognita* is distributed into southern California, Arizona, and New Mexico. Similar distribution patterns are recognized throughout the world, and producers should be knowledgeable about which nematodes are present in their production areas in order to most effectively manage the pest.

The second-stage of the root-knot nematode larvae emerge from the egg, travel through the soil water film to the host plant root or tuber, and enter through natural openings or cracks in the cuticle or periderm. Once in the plant, the larvae are able to move to the vascular tissue where they remain. The larvae cause the plant to produce giant cells which appear as galls on root or tuber surfaces. These cells then supply food for the larvae. The females become 1-2 mm (0.04-0.08 inches) long and pear-shaped while the males become wormlike, 1-1.5 mm (0.04-0.06 inches) long, and leave the root or tuber. The females produce up to 1000 eggs which hatch to release young larvae. The generation cycle in nematode species is dependent on temperature, but will be complete in 20 to 60 days.

The symptoms of nematode damage are most apparent on mature tubers or on roots. The prominent knots produced by some *Meloidogyne* species are one diagnostic symptom, but may not be obvious with species such as *M. chitwoodi*. External tuber symptoms may be mild with some infections, but symptoms below the peel can still be extensive. When suspect tubers are lightly peeled, the swollen females can be seen. They will be white or cream color early in the season, but brown, necrotic cells will develop around them as the tuber matures. These brown spots create a quality problem and are counted as defects in both fresh and processed potato grading.

Control of root-knot nematodes can be achieved by using nematicides, planting only early potato cultivars, and with some crop rotations. The best management practice may include all three of the above measures plus selection of seed from nematode-free areas, and through practices of sanitation and fallow cultivation. Since *M. chitwoodi* can increase in storage, potatoes that are infected should be processed as soon as possible.

Current control measures

   a.
   b.
   c.

Root-Lesion Nematode (*Pratylenchus neglectus, P. penetrans, P. crenatus, P. brachyurus, P. scribneri*). The general life cycle of the root-lesion nematode is the same as the root-knot nematode. The primary difference between these nematodes is that the plant does not produce giant cells when infected by the lesion nematode. The larvae enter the root just behind the root tip and then move through the roots. The lesions caused by this nematode are sites for secondary infection by other pathogens such as *Fusarium* spp., *Verticillium* spp., and bacteria. Tubers can be damaged by *P. brachyurus* or *P. scribneri*, but these are not a problem in the United States. Control can be achieved with fumigant or nonfumigant nematicides. Rotation with other crops is usually not effective because of the wide host range of the lesion nematode.

Current control measures

a.

b.

c.

Stubby-Root Nematode (*Paratrichordorus pachydermus, P. christiei,* and *Trichodorus primitivus.* Control of the stubby-root nematode is a problem in potato production because it transmits the tobacco rattle virus which causes corky ringspot (see Viruses section earlier in this chapter). This nematode hatches in the soil and feeds on roots without entering root tissue like the root-knot or lesion nematodes. Their life cycle is about 45 days at $15°$-$20°C$ ($59°$-$68°F$).

This nematode feeds on a number of hosts and moves deeply in the soil profile, which makes control somewhat difficult. Fields that have produced potatoes infected with tobacco rattle virus should be avoided. Fumigants and nonfumigant nematicides have been effective but caution must be used because of the mobility of the nematode.

Nematodes can cause severe damage both in terms of yield and quality. Symptoms are generally not seen on above-ground plant parts, although general lack of vigor or yellowing and wilting can occur. Preventative control measures are more effective than remedial measures and are usually more cost effective. Use certified potato seed from areas that are not suspected of having nematodes. Fumigate or use non-fumigant nematicides when fields are known to have nematodes. The interaction of population dynamics of each nematode species with any specific climate makes it difficult to generalize when it is necessary to apply control measures. The assessment of each situation should be made by collecting soil population data and consulting with experts in the area of production.

Current control measures

a.

b.

c.

SUMMARY

Disease may be present in the soil, carried into the field in water, by implements, insects or wind, or may be introduced in the seed. Although high-quality, low disease-content seed may be more expensive to purchase, its benefit of reduced disease in the field planting may result in less additional input cost during other phases of the growth or storage of the crop.

The onset of a disease can be very rapid; therefore, producers should monitor fields, storages, and soil conditions frequently. A consistent monitoring program and collection of a few pieces of data during each observation period can be invaluable tools to maintain a productive potato crop. The **Appendix** includes a field monitoring form to help with data collection for a good field evaluation.

Chapter 8

Physiological Disorders

INTRODUCTION

Potatoes are subject to various disorders that are not caused by diseases, insects, or other pests. These are commonly referred to as physiological disorders and are usually the result either of an environmental factor causing abnormal growth patterns, or of biochemical events. Potato cultivars differ in their susceptibility to the various disorders, and some disorders do not occur in some growing areas. Many times, the cause of physiological disorders is poorly understood and therefore control measures are oftentimes vague. The following are the most common physiological disorders.

TYPES OF DISORDERS

Malformations. Some cultivars become irregular in shape when stressed for water or nutrients. The tubers may be generally rough in appearance, have knobs (actually tuber branches), be pointed on one end or the other, or shaped like a dumbbell. These malformations usually result from uneven application of water (or untimely rains). As mentioned earlier, water stresses can be common even when the field is well watered if the water-holding capacity of the soil is low and water cannot be supplied at frequent enough intervals. When a stress occurs, cell division may be irreversibly stopped, perhaps from a hormonal signal, and therefore the final size potential of the tuber will have been reduced. Because the tuber is really just an underground stem, when the stress is relieved, new growth will occur most frequently at the areas with meristems or active cell division. These areas are at the apical (bud) end of the tuber or at the nodes (eyes). That is why when knobs and growth abnormalities of this kind occur, they are found in these locations and not randomly around the tuber surface. The date or approximate time of the season that a stress is incurred can be surmised by the location of constrictions of the tuber surface (Iritani, 1981).

Control of malformations is best achieved by selecting resistant cultivars. Russet Burbank is very susceptible to this disorder, whereas cultivars such as Norkotah are somewhat resistant. The number of malformations will be reduced when soil moisture can be maintained at a level which prevents the plant from becoming stressed. Large applications of nitrogen fertilizer during tuber bulking have also been implied to cause this disorder. The effect of nitrogen may be to increase the leaf area, which increases total transpiration of the plant, and therefore makes it more subject to water stress. Adjustment of the leaf canopy through proper cultivar selection and fertilization programs will reduce malformations.

Growth Cracks. Growth cracks are splits in the tuber surface which usually penetrate through the cortex and into the perimedullary tissue. These are different from the cracks that occur at harvest when tubers are turgid and are impacted by dropping them onto a hard surface (thumbnail checks or shatter bruise). The large growth cracks which occur during tuber bulking are presumed to be caused by irregular water availability (stress) and are more prevalent with widely spaced plants.

Hollow Heart/Brown Center. The two disorders referred to as hollow heart and brown center may be related, and often occur under similar conditions. Brown center is manifested as a group of brown (dead) cells found in the pith or central area of the tuber. The brown area may be very small; the cells may be quite disperse or very dense, dark, and concentrated. Brown center occurs during early tuber growth and is thought to be related to water stress within the plant. Hiller (in Li, 1985) has shown that its occurrence can be increased by subjecting the root zone to low temperature ($< 10°C$ ($< 50°F$)) during the tuber initiation period, when tubers are about 2.5-3.8 cm (1-1 1/2 in) in diameter. He also suggests that high moisture levels and rapid growth rates increase brown center occurrence.

Hollow heart may be related to brown center in that it could conceivably be the most severe manifestation of this type of stress disorder. Hollow heart is a cavity or series of cavities found in the center of the tuber (pith area) and may be located at one or more locations along the length of the tuber. Multiple hollow heart cavities may be connected by brown cells such as those found in the brown center disorder. Some individual hollow heart cavities may be surrounded by these same brown oxidized cells. Hollow heart also seems to be related to relative tuber growth rate and stress at some point in the tuber growth period. No control is known but increased plant population and reduced stress from water are suggested to moderate the problem. The largest tubers at harvest are most frequently the tubers containing these disorders.

Sugar End, Translucent End, Jelly End. These disorders are each different manifestations of the same disorder with differing degrees of severity. They have been researched more than any of the other disorders and probably have been studied more than all the rest put together.

In order to understand the nature of these problems, we must understand a basic principle in potato physiology. The tuber begins to grow as a result of hormone changes in the foliage, roots, and stolons (discussed under Tuberization in Chapter 4). After growth begins, starch, proteins, and other materials accumulate in the tubers.

During tuber growth, sugar, amino acids, and other basic building blocks are transported from the foliage to the tuber where starch and proteins are rapidly manufactured. At this point, the tuber is called a sink, that is, it provides a location for the leaves (source) to unload all of its products (sugars and amino acids). This relationship of source and sink is a competitive one particularly during early tuber growth, and can be altered by growth hormones (see Chapter 4), the climate (photoperiod), and stress conditions (nitrogen or water).

If the foliage becomes stressed for water and its ability to produce sugars is reduced and/or its hormone production (ABA, GA) is altered, changes in growth should be expected. It is well known that when plants are stressed, ABA levels generally increase and stresses, such as for nitrogen, can increase ABA levels in potatoes, resulting in early tuber formation (see Foliage Growth Tuberization sections in Chapter 4). When plant stress is relieved, renewed growth may occur as the plant shifts back to its original growth pattern as a result of hormone changes.

Stress due to irregular water availability is known to be the primary cause of sugar end, translucent end, or jelly end. Many research reports have proved beyond a doubt that withholding water until the plant is stressed and then applying it will induce these disorders. The nature of the disorder results from the plant (foliage) sending a signal to the tubers, which causes the starch to be hydrolyzed (broken down) to sugars, which can be used by the foliage to meet its demands (the tuber may produce the signal as well). In this instance, the tuber may be functioning like the seed piece that was planted in the spring. It hydrolyzes the starch to sugars (glucose) which may be transported to the top. Only, in this case, the transport of the sugar to the top is not complete and depending on the stress level, and the amount of starch hydrolyzed and sugar remaining in the tuber will vary. A mild sugar end will result from a low stress level, and a translucent end tuber will result if the stress is great enough to induce most, if not all, of the starch to be degraded (resulting in the translucent appearance). The amount of sugar left will depend on the amount of starch hydrolyzed and the amount of sugar transported or metabolized by other means. Jelly ends appear when water, as well as starch, are removed from the stem end of the tuber and therefore the end shrivels and drys leaving only the periderm. Entire tubers (usually small) may be reabsorbed in this fashion (Dean, observation). This disorder is very important in the processing industry. Potatoes with these disorders usually process poorly because of their high reducing sugar and low starch content.

Control of these disorders can be achieved by avoiding stress and by planting resistant cultivars. It is most important to reduce fluctuations in soil moisture such as are common with furrow irrigation. Maintain soil moisture at a uniform optimum level for the soil type.

Internal Brown Spot/Heat Necrosis. Internal brown spot (IBS) and heat necrosis have also been called by other names, and confusion exists when different names are used. The disorders are described as having necrotic (dead, oxidized) cells dispersed in the perimedullary tissue of the tuber, but sometimes located in the cortex or primarily on one end (frequently the apical end) of the tuber. Sometimes IBS is identified as having fairly dense areas of dead cells between the vascular tissue and the periderm. Sometimes IBS is differentiated from heat necrosis, where IBS refers to the diffuse, cell-speckling that occurs, and heat necrosis refers to the manifestation of the dead cell mass in the cortical region.

IBS-heat necrosis is usually first observed during rapid tuber growth periods, but has been found in small tubers after they were stored (Thornton, unpublished). At harvest, the largest tubers usually have more of the disorder than smaller tubers from the same plant. This is one of the few examples of a physiological disorder that appears to increase during storage.

Some cultivars are susceptible to IBS/heat necrosis while others are not. Russet Burbank and Atlantic appear to be relatively susceptible. Several factors of this disorder have suggested to researchers that this may be a pathological disease rather than a disorder. The incidence of this disorder has increased during the past ten years. It increases in severity during long-term storage, and it does not seem to be particularly associated with plant stresses. A condition similar to IBS has been induced by calcium deficiency in controlled conditions (Li, 1985) and by phosphorous deficiency (Houghland, 1949). Calcium applications have not been effective in controlling the disorder.

Enlarged Lenticels. Lenticels are the pores in the skin (periderm) through which gas exchange is maintained by potato tubers. Although the periderm itself is permeable to gases, the lenticels are apparently necessary for this function also. When soil moisture becomes high enough to cause a film of water to cover the surface of the tuber, the lenticels will enlarge. They become pronounced white spots covering the tuber surface. These white spots are actually a proliferation of cells from the lenticels. These enlarged masses of cells may provide entry ports for diseases, particularly bacteria, because they are not suberized.

Soil moisture must be kept below saturation for this disorder to be avoided. Careful monitoring of the lenticels can provide a good indicator that field overwatering has occurred. If enlarged lenticels are found, irrigation should be reduced.

Blackheart. This disorder also results from overwatering but may be caused by overheating as well. Blackheart is identified by the black tissue found in the center of the tuber when it is cut. It results from conditions that cause the tuber to become anaerobic. These include saturated soil conditions, overheating in storage, transit, or in the field; or reduced oxygen in storage as a result of poor ventilation.

Maintain proper soil moisture, do not expose tubers to high temperatures, and ventilate storage areas properly to prevent this disorder.

Heat Sprouts. Some cultivars are susceptible to sprouting under extreme temperature conditions in the field. Depending on the time during the growth cycle, heat sprouting may be the formation of new stolons and/or shoots from small tubers, to the sprouting of several buds from several eyes on a mature tuber prior to harvest. The latter disorder may be more prevalent on giant hill plants.

Controlling this disorder may only be by cultivar selection. Climatic conditions that cause heat sprouting conditions are not frequent in the major potato growing regions. Leaving tubers in the field after the foliage has died may increase the severity of the disorder. Select seed lots that have a low percentage of giant hill plants.

Giant Hill. Giant hill is a somatic mutation that is propagated within the seed source. When plants are selected for seed increase, most seed producers avoid plants that have giant hill characteristics. Some producers have selected giant hill plants to produce new cultivars with the belief that the disease resistance inherent in them will be beneficial.

The disorder produces plants that usually bloom over a longer period of time, have a single-course stem, a rough leaf texture, produce branches in the leaf axils, and do not die when the majority of other normal plants die. The roots of giant hill plants stay healthy very late in the season, and the tubers may be attached with large stolons. The tubers can be smooth but are usually rough, knobby, have deep eyes, and sprout in the field if stressed.

If giant hill is expressed in the growing area, seed lots should be selected that have minimal giant hill. Russet Burbank seed lots have been observed by the author to have 1-1.5 percent giant hill. Seed fields can be rogued for this disorder by keeping fertility levels low and roguing late in the season. Late-maturing plants or those with predominate stems or lateral branches should be removed.

Elephant Hide. Elephant hide is identified by the rough irregular surface and thick periderm on portions of the affected tuber. The tubers are difficult to peel during processing because of the periderm thickening. No cause has been definitely identified although high fertilizer salts placed next to the tuber has been suggested.

Greening. Tubers will turn green when exposed to light. Any conditions which expose tubers to natural or artificial light for even short periods may result in the synthesis of chlorophyll in the cells just beneath the skin. Alkaloids are often associated with this greening and because of their toxicity, it is advised that green tubers not be consumed.

Form wide hills so that the ends of tubers do not extend out of the soil, and prevent soil cracking that exposes tubers to light. Do not leave windrowed potatoes in the field for extended periods. Keep lights off in storage areas.

Frost Damage. This environmental injury occurs frequently in some growing regions when temperatures drop below $-2°C$ ($28°F$). Frosted tubers first become wet on the surface and may then shrivel slightly where the damage occurred. The tuber surface will often turn a silvery brown color after a period in storage, and the damaged tissue will slough off. The damaged tissue is clearly differentiated from undamaged tissue, and no odor will be present until infection by pathogens occurs.

If the weather will permit frosted potatoes to remain in the field for two to five days prior to harvest, the damaged tissues will usually slough off in the field. Damaged tubers will then be easy to grade out either on the harvester or going into storage. Some seed growers have washed potatoes going into storage with large volumes of fresh (not recycled) water and have been successful in storing frosted seed. Storages must be well ventilated to dry the tuber surfaces, or rot will occur. Washing tubers before they go into storage is not recommended, but may be considered as a salvage operation.

Stem End Browning (discoloration). This disorder occurs under many circumstances and in different orders of severity. It is defined as having brown streaks on the stem end of the tuber, and which are restricted to the vascular tissue. It may be confused with net necrosis caused by the leafroll virus. The causes seem to be quite varied but are associated with vine death. Vine killing with chemicals, particularly under stressed conditions, is one possible cause, but the disorder is also seen in tubers from vines killed by frost, from vines that died a natural death, or from vines that have been physically removed.

A definite control measure has not been found.

Bruise (blackspot, thumbnail check, shatter). These are the three primary manifestations of damage to potato tubers that occur during harvest. The most important damage is called blackspot, followed by thumbnail crack and then shatter bruise. Each of these are distinct disorders but are caused by impacts during harvesting and handling.

When the tuber is carefully removed from the soil, no thumbnail cracks, shatter bruises, or blackspots can be found. However, after harvesting and handling the tubers mechanically, and depending on the cultivar, the temperature, and many other factors, these disorders will appear.

Shatter bruise is a fracturing of the tuber surface. Tissue beneath the periderm that usually remains clear, but some darkened fractures may be present. It is caused by an impact to the surface of the tuber.

Thumbnail checks (air checks) are caused by slight impacts to the surface. They are sometimes associated with small blackspots just beneath the periderm (Dean, 1989b). The damages appear as semicircles or as if the tuber had been gouged with a thumbnail (hence the name).

Blackspot bruises are damaged areas in the flesh just beneath the skin that turn brown, grey, or black after six to 24 hours following an impact. In this case, the cells may not be fractured, but the cell integrity has been damaged so that its internal compartmentation has been altered. This results in an enzyme (polyphenoloxidase) coming in contact with a substrate (primarily the amino acid tyrosine) and producing a black pigment called melanin. The amount of blackspot that occurs depends on how much tyrosine and polyphenoloxidase there is in the cell solution at the time of the injury.

Blackspot has been shown to be more severe when potassium nutrition is poor (Mulder, 1949; Kunkel, 1969). Some cultivars are very susceptible (Lemhi Russet), others are fairly resistant (Atlantic), and cultivars such as Russet Burbank are intermediate. Potatoes harvested under cold temperatures have been shown to be more susceptible to blackspot (Smittle et al., 1974), and more turgid tubers are relatively resistant to blackspot but more susceptible to shatter bruise.

These disorders can be reduced by maintaining good soil potassium levels, adequate soil moisture at harvest, and minimizing damage (see Harvesting and Handling, and Harvester Operation sections in Chapter 9).

Pressure Bruise. Pressure bruise is a grey to black area under the skin of the tuber that results from pressure being applied to the surface during storage under dehydrating conditions. A portion of the tuber turns grey to black and the surface is usually compressed. Avoid low humidity storage conditions.

Chapter 9

Harvesting, Handling, and Storage

HARVESTING AND HANDLING

The date of harvesting is dependent upon the intended use of the crop. If the crop is destined for fresh market, three of the major considerations determining the timing of the harvest are tuber size, shape, and appearance. If the tubers are to be processed into french fried, molded, or dehydrated products, compositional factors are as important as shape or appearance.

Once the potato crop has reached the stage where economic returns or quality dictate that harvest should commence, vines may be killed with non-translocated herbicides, or beaten off with some type of vine removal device. There are several advantages to killing the vines prior to harvest including: (1) allowing the development of a firm periderm which decreases scuffing of the skin, (2) facilitating harvesting, (3) controlling the spread of diseases such as late blight, and (4) controlling tuber size. This process is usually done ten days to two weeks prior to harvest. The tubers can also be harvested from plants that are still green, or from plants that have died from disease, insect damage, low fertility, or climatic conditions. There is no single accepted set of practices which cover all growing areas. However, when possible, soil moisture should be maintained to prevent desiccation of the tubers. Adequate aeration should be established to prevent disease problems, such as those associated with bacterial rots. The soil moisture content desired will depend on the soil type and its water-holding capacity, climate during the harvest period, and control or lack of control of water application.

Yield can be estimated periodically in order to prioritize fields for digging. If a field is still growing it may be saved for a later harvest, while one that has a reduced growth rate due to maturity, disease infection, or other causes can be harvested earlier. Select three areas within the field that are representative of the average conditions and harvest two 3-meter (ten-foot) sections adjacent to each other in each area. Weigh the entire sample. The yield can be approximated from the following table:

Row Spacing	Total Sample Weight	Conversion × Factor	=	Yield/acre (lbs)	Yield/hectare kg
	lbs kg				
32 inches	_____	× 272.3	=	_____	_____
81 cm		305.0			
34 inches	_____	× 266.2	=	_____	_____
86 cm		298.1			
36 inches	_____	× 242.0	=	_____	_____
91 cm		271.0			

Find the row spacing appropriate for the field, enter the total sample weight for the three 6-meter (20-foot) samples, multiply by the conversion factor to obtain a kg per hectare or a pound per acre figure. The yield sample can also be used for tuber quality determinations.

HARVESTER OPERATION

The physical condition of the tubers is a major consideration, regardless of the destination of the potatoes, because of its relationship to bruising. When a potato tuber is struck by an object, a bruise may appear within six to 24 hours. It is important to understand some principles of harvester operation in order to avoid blackspot, shatter, or thumbnail check damage.

The mechanics of harvesting revolve around the operation of a machine which consists of a digging blade, and a series of chains which convey the potatoes into a truck or deposit them on top of the ground in windrows (Figure 16). The harvester may be self-propelled or be pulled by some other power unit (e.g., tractor). The digging blade is placed into the soil to a depth of about 20 cm (8 in) and as the harvester moves forward, the tubers are conveyed to the

FIGURE 16. A pull-type potato harvester.

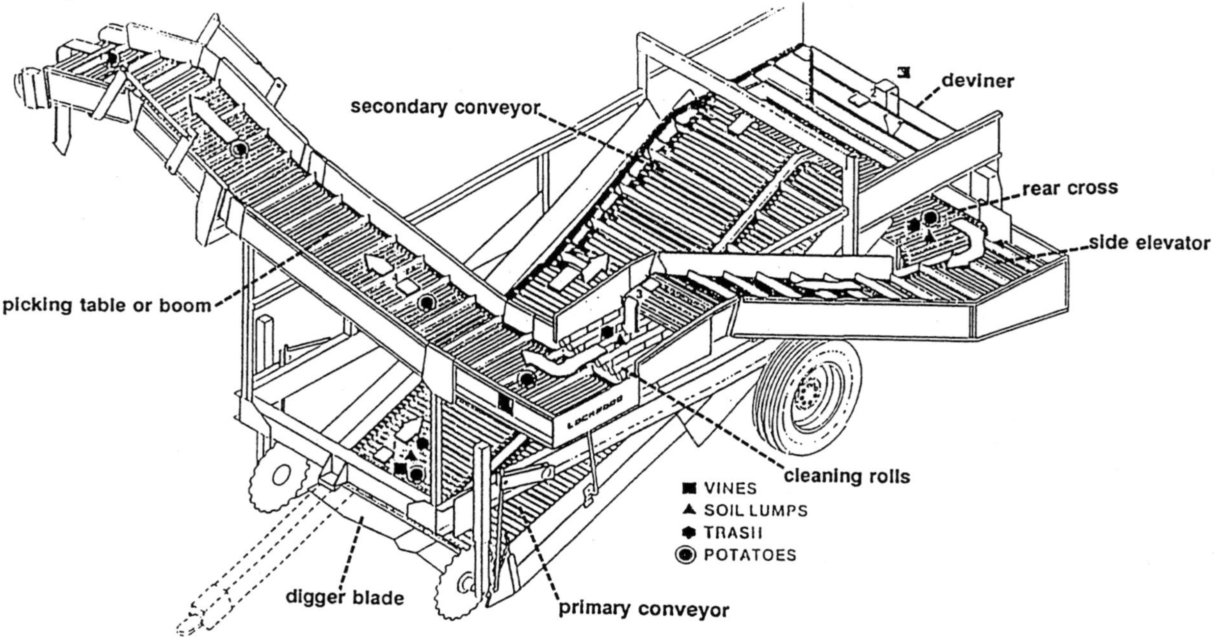

deviner

rear cross

side elevator

■ devlner

secondary conveyor

picking table or boom

cleaning rolls

■ VINES
▲ SOIL LUMPS
♦ TRASII
◉ POTATOES

primary conveyor

digger blade

Potato harvester by Lockwood Corp.

primary chain located behind the blade. The speed at which the harvester travels and the speed at which the chains operate depend on the soil type, and condition and the yield of tubers in the field (Thornton, Smittle, and Peterson, 1974). The soil type and condition dictates the operating speed by the drag that is caused on the machine, by the amount of rock which needs to be removed (mechanically or by hand), and by the ease in which the soil can be eliminated from the chains. The more potatoes present per unit land area, the more tubers which are going to be present on the harvester at a given time and at a given speed. The chains on the harvester should be filled with potatoes so that tubers do not bounce or roll on the chains. In order to fill the harvester chains, the speed of the chains must be matched to the ground speed of the harvester, which is influenced by the yield of tubers and soil conditions.

Adjustment of the Harvester

Proper adjustment of the harvester will accommodate rapid harvest and minimize bruise damage. The chain speed to ground speed ratios can be measured and adjusted by simple sprocket changes. The following techniques can be used to adjust the harvester:

1. Accurately measure the forward speed of the harvester by timing its travel over a given distance. Note the rpm of the power take off (PTO) shaft.
2. Remove harvester chain guards covering the shafts to which chain drive sprockets are attached.
3. With the PTO at normal operating speed and the tractor operating at normal forward speed, measure the rpms of each drive sprocket shaft. (This must be done while the harvester is digging.)
4. After stopping the harvester, measure the chain pitch (distance between chain rods from center to center) and the number of teeth on the drive sprocket.

Convert shaft rpm to mph using the following formula:

rpm × chain pitch (in)/12 × number of teeth on drive sprocket
= ft/min

$$\text{rpm} \times \frac{\text{chain pitch}}{12} \times \text{teeth} = \text{ft/min}$$

$$\text{ft/min} \div 88 = \text{mph}.$$

Example: primary chain drive shaft speed = 133 rpm.

rpm × chain pitch (cm)/100 × number of teeth on the drive
   sprocket
= meters/min. meters/min ÷ 16.7 = KM/hour.

$$133^{\text{rpm}} \times \frac{1.76^{\text{in}}}{12 \text{ in/ft}} \times 12^{\text{teeth}} = 234.1 \text{ ft/min}$$

234.1 ft/min ÷ 88 = 2.66 mph (speed of primary chain)

If the measured ground speed was 2.2 mph, then the primary chain-to-ground speed ratio is:

$$2.66 \div 2.2 = 1.21$$

The ratio of chain speeds to ground speed should approximate those in Tables 18 and 19.

If the ratio is found to be different than desired, the speed of the chain can be altered by changing the driver or driven sprockets to the appropriate size. It should be noted that some sprockets drive more than one chain directly or indirectly. Therefore, care must be taken to inadvertently avoid changing the speed of chains which have already been correctly set up.

STORAGE

Most of the late crop of potatoes in the United States is stored for periods of three to nine months. During storage and marketing of potatoes, disease, sprouting, water loss, sugar buildup, and greening should be prevented. Badly bruised and diseased potatoes should not be placed in storage.

TABLE 18. Harvester chain/forward speed ratios for heavy soils.

	Yield (Cwt/Acre)						
	(100*)	(200*)	(300*)	(400)	(500)	(600)	(700)
Primary	1.1	1.1	1.1	1.1	1.1	1.1	1.1
Secondary	.7	.7	.7	.7	.7	.7	.7
Rear Cross & Elevator	.2	.3	.4	.5	.5	.6	.6
Boom	.2	.2	.3	.4	.4	.5	.6

*The ratios at these yield levels have not been adequately tested and therefore must be considered theoretical values.
Thornton, Smittle, and Peterson, 1974
Cwt/A × 1.12 = quintals/hectare

TABLE 19. Harvester chain/forward speed ratios for sandy soils.

	Yield (Cwt/Acre)						
	(100*)	(200*)	(300*)	(400)	(500)	(600)	(700)
Primary	.9	.9	.9	.9	.9	.9	.9
Secondary	.6	.6	.6	.6	.6	.6	.6
Rear Cross & Elevator	.2	.3	.4	.5	.5	.6	.7
Boom	.2	.2	.3	.4	.4	.5	.6

*The ratios at these yield levels have not been adequately tested and therefore must be considered theoretical values.
Thornton, Smittle, and Peterson, 1974
Cwt/Acre × 1.12 = quintals/hectare

Temperature

Temperature control is essential for long-term storage of potatoes. Potatoes must be: (1) cooled to remove field heat after being loaded into the storage; (2) provided adequate temperatures for wounds to heal, and (3) provided optimal temperatures for long-term storage depending on the anticipated use of the crop.

The first requirement for cooling capacity in storage is for the removal of field heat (the amount of heat in the potatoes from the field). If potatoes are harvested during cold weather when tubers are below 10°C (50°F), this is not a significant factor. The second need for cooling is to remove the heat caused by respiration. When potatoes are immature or after they are injured, they have high respiration rates. The respiration generates heat that must be removed. Depending on the combined effects of these two heat sources, cooling requirements can be determined. Cooling can be achieved by using standard refrigeration units, evaporative cooling units, or by using ambient air when the temperature is satisfactory.

Because potatoes are damaged during the harvesting and handling operations, these injuries must be healed in order to achieve long-term storage with minimal losses. The process of wound healing involves the reforming of a new skin which contains a waterproofing material called suberin (Kolattukudy and Dean, 1974). This process takes approximately ten days at 22°C (70°F) and is very slow at reduced temperatures (Dean, 1989a). Potatoes should be held at temperatures above 10°C (50°F) for at least two weeks prior to lowering the temperature to the desired long-term storage temperature. The cooling air should be maintained at 1°-2°C (0.5°-1.0°F) cooler than the tuber temperature during the time that the storage temperature is being lowered. If large temperature differences are used, dehydration of the tubers may occur because of the vapor pressure deficit that is created. The temperature should be lowered slowly over two to four weeks. Potatoes intended for processing, particularly those intended for production of potato chips or french fries, should not be stored at low temperatures (< 5°C (< 45°F)) or there will be a buildup of reducing sugars in the tubers. The lower limit is dependent upon the cultivar being stored and the planned use for it. Therefore, local recommendations in each processor's area should be determined prior to storage. High reducing sugars and the presence of certain amino acids result in very dark chips or fries being produced during the frying process. Sugars can be reconverted to starch or respired off prior to processing by placing the tubers at 20°-22°C (68°-72°F) for one to four weeks (a process called reconditioning). However, not all of the sugar is always reconverted to starch and it is often difficult to predict just how successful reconditioning will be.

Potatoes intended for table stock or "seed" should be held at 3°-5°C (37°-42°F) for long-term storage in order to prevent sprouting and to help decrease disease incidence, both of which are inhibited by low temperature.

Seed potatoes should be warmed for five to ten days prior to planting. If no sprouting is evident, seed potatoes can be exposed to temperatures of 12.5°-20°C (55°-68°F) for ten days to two weeks to induce sprouting.

Humidity

The potato contains 75 to 80 percent moisture which must be maintained to prevent shriveling (shrinking). The relative humidity should be kept at 95 to 100 percent to prevent excessive water loss. Any relative humidity below 100 percent will result in water loss from the tubers. A high quality storage operation usually expects a 4 to 7 percent loss.

Ventilation

Potato storages should be equipped with ventilation systems that provide for the exchange of the entire volume of air frequently enough to maintain adequate oxygen levels and to provide necessary cooling. Air velocities of approximately 4.8 hectoliter (hl)/min (17 cfm per ton) are recommended to achieve initial storage conditions. After the potatoes are at the desired temperature, the air velocity may be reduced to (2 or 2.3 hl/min (7 or 8 cfm per ton) for holding (Potato Storage and Ventilation, WSU EM2799). Low levels of oxygen will enhance bacterial decay and cause blackheart of the tubers. Maintain air movement above the pile with additional fans to minimize condensation on the building structure.

Sprout Control

The higher the storage temperature, the more quickly tubers will sprout. There is a limit, however, as to how low a temperature tubers can be held at because of the potential for sugar buildup and chilling injury. Potatoes held at 7°C (45°F) for long-term storage should be treated with a sprout inhibitor, but seed potatoes should never be treated.

CIPC (isopropyl-[3-chlorophenyl] carbamate) is one material used as a sprout inhibitor. It is usually applied after the potatoes have been in storage for several weeks and after wound healing has occurred, because CIPC will interfere with this process. It is usually added as an aerosol through the ventilation system. It may also be applied as a dip or added to the wash water of potatoes coming out of storage destined for fresh market. Application of too low a concentration of CIPC may cause the development of internal sprouts (sprouts that grow into the tuber).

Maleic hydrazide is used as a foliar spray during the growing season to prevent sprouting in storage. It is applied after the desired number of tubers have reached at least two inches in diameter.

Irradiation of tubers with gamma rays also inhibits sprouting but may also inhibit wound healing. The difficulties of moving the potatoes into storage for curing and then moving them again to be irradiated, plus the volume of tubers to be handled and the costs involved, have prevented irradiation from being used commercially.

Current sprout control measures

    a.

    b.

    c.

Chapter 10

Potato Quality

INTRODUCTION

There are several tuber quality factors that dictate the value of the potato crop and which growers should be familiar with. Some of these factors are influenced by the grower and others are affected by the environment.

GRADE AND SIZE

The grade and size of the tubers is influenced by almost every decision in managing the production system. From variety selection to spacing, fertility, irrigation, and storage, the grade quality can be changed at any point. There are three USDA standards for potatoes: (1) fresh potatoes; (2) processed potatoes; and (3) seed potatoes. A copy of the standards can be obtained from the United States Department of Agriculture representative in any potato growing area. A summary of the fresh potato standard is provided in Table 20; the definitions of the terms used are presented alphabetically in the following section.

It should be understood that the USDA grades are minimum standards and do not dictate the industry standards except in that regard. Individual states, regions, and private companies have developed standards which may exceed these minimums. The final consumer will ultimately establish what grade is acceptable through purchasing practices in retail outlets.

TABLE 20. Summary of grades of potatoes for fresh pack.

	U.S. Extra No. 1	U.S. No. 1	U.S. Commercial	U.S. No 2
1. Varietal Characteristics	Similar	Similar	Similar	Similar
2. Firmness	Firm	Firm	Firm	
3. Cleanliness	Clean	Fairly clean	Not seriously damaged by dirt	Not seriously damaged by dirt
4. Maturity 1. USDA	Fairly well matured	No spec.	No spec.	No spec.
5. Shape	Fairly well shaped with 50% well shaped	Fairly well shaped	Fairly well shaped	Not seriously misshapen
6. Maximum Defects				
a.	2%	3%	3%	3%
Freezing, blackheart, southern bacterial wilt, ring rot, late blight, soft rot, or wet breakdown (maximum)				
b.	0.5%	1%	1%	1%
Frozen, soft rot, and wet breakdown (maximum)				
c. Internals (maximum):				
i. "Injury"	5%	–	–	–
ii. "Damage"	–	5%	20%	–
iii. "Serious damage"	–	–	6%	6%

146

	U.S. Extra No. 1	U.S. No. 1	U.S. Commercial	U.S. No. 2
d. Externals (maximum)				
i. "Damage"	5%	5%	20%	0
ii. "Serious Damage"	–		6%	6%
e. Total of all defects (maximum) (6.a + 6.c. + 6.d.)	5%	8%	20% all 10% "Serious"	10% (Serious)
7. Minimum size				
USDA	2 1/4" or 5 oz.	1 7/8"	1 7/8"	1 1/2"

USDA POTATO STANDARDS TERMS

Clean: at least 90 percent of the potatoes in any lot are practically free from dirt or staining, and practically no loose dirt or other foreign matter is present in the container.

Damage: any defect, or any combination of defects, which materially detract from the edible or marketing quality, or the internal or external appearance of the potato; or any external defect which cannot be removed without a loss of more than 5 percent of the total weight of the potato.

External defects: defects which can be detected externally; however, cutting may be required to determine the extent of the injury.

Fairly clean: at least 90 percent of the potatoes in any lot are reasonably free from dirt or staining, and not more than a slight amount of loose dirt or foreign matter is present in the container.

Fairly well-matured: the skins of the potatoes are generally, fairly firmly-set, and not more than 10 percent of the potatoes in the lot have more than one-fourth of the skin missing or "feathered."

Fairly well-shaped: the potato is not materially pointed, dumbbell-shaped, or otherwise materially deformed.

Firm: the potato is not shriveled or flabby.

Freezing: the potato is frozen or shows evidence of having been frozen.

Injury: any defect, or any combination of defects, which more than slightly detracts from the edible or marketing quality, or the internal or external appearances of the potato; or any internal defect outside of or not entirely confined within the vascular ring, which cannot be removed without a loss of more than 3 percent of the total weight of the potato.

Mature: the skins of the potatoes are generally firmly set, and not more than 5 percent of the potatoes in the lot have more than one-tenth of the skin missing or "feathered."

Serious damage: any defect, or any combination of defects, which seriously detracts from the edible or marketing quality, or the internal or external appearances of the potato; or any external defect which cannot be removed without a loss of more than 10 percent of the total weight of the potato.

Seriously misshapen: the potato is seriously pointed, dumbbell-shaped, or otherwise badly deformed.

Similar varietal characteristics: the potatoes in any lot have the same general shape, color and character of skin, and color of flesh.

Slightly skinned: not more than 10 percent of the potatoes in the lot have more than one-fourth of the skin missing or "feathered."

Soft rot or wet breakdown: any soft, mushy, or leaky condition of the tissue such as slimy soft rot, leak, or wet breakdown following freezing injury.

Well shaped: the potato has the normal shape for the variety.

SPECIFIC GRAVITY

Grading for processing is usually similar but less restrictive on size and shape. In addition to size and appearance grade standards, processors may grade potatoes based on the amount of starch present in the tuber measured by specific gravity tests. Specific gravity is an important raw product quality characteristic, and therefore should be explained in some detail.

What Is Specific Gravity?

Specific gravity is the density of a potato tuber or any other object relative to the density of water at a given temperature. At 10°C (50°F), the specific gravity of water is 1.000. When placed in a container of water, a potato tuber will sink because it has a specific gravity greater than 1.000. The non-water components (dry matter or solids) of the tuber result in a tuber with a given specific gravity. The major dry matter components of the potato tuber are made up of carbohydrates, primarily starch (80-85 percent starch plus 10-15 percent cellulose and 1-5 percent soluble sugars); therefore, the specific gravity of the tuber mainly reflects the amount of starch present.

The term specific gravity is used as an indirect measure of the non-water components of potato tubers. These non-water components are the dry matter or solids of the tuber, primarily starch. Therefore, factors which affect starch level affect dry matter or solids content, and thus, affect specific gravity.

TABLE 21. The solids content (%) in 3 parts of tubers of 5 potato cultivars.

Tuber portion	White Rose	Red La Soda	Kennebec	Russet Burbank	Norchip
Bud	16.1	18.8	18.3	17.5	19.7
Pith	14.3	13.6	14.0	15/0	15.0
Stem	21.2	23.5	23.4	22.1	23.4

Weaver, M. L., H. Timm, M. Nonaka, R. N. Sayre, R. M. Reeve, R. M. McCready and L. C. Whitehand. 1978. Potato Composition I: Tissue selection and its effects on solids content and amylose/amylopectin ratios. *Am. Pot. J.* 55:73-82.

The concentration of starch also varies from one end of the tuber to the other in many potato cultivars (Table 21). Potato tubers from the same field and from individual plants within the same field also exhibit variability in starch content. Much of the variability between tubers within a lot and from individual plants can be accounted for by the variability in tuber size. The highest starch level that tubers can reach appears to be under genetic control, as evidenced by the differences among cultivars (Table 22).

To produce quality processed potato products, tubers must have a high specific gravity (starch content) and low soluble sugars. Potato tubers with medium to high specific gravity are generally preferred for baking. Tubers with a lower specific gravity are used satisfactorily for home frying, fresh processing, salads, and mashed potatoes. However, individuals differ in their preferences for specific gravity levels for the various uses, especially in the home.

Why Is Specific Gravity Important?

As discussed, the specific gravity of the tuber is indicative of its starch content and total dry matter. Although there are many variables involved in the cooking quality of potatoes, the best correlations found to date are those of dry matter, specific gravity, and starch grain size. There is also a high correlation between the size of individual starch grains and the specific gravity, of the tuber (Sharma and Thompson, 1956).

Many studies have been performed to determine how specific gravity affects the cooking quality of potatoes. It is known that higher

TABLE 22. Varietal differences and range in specific gravity within cultivars of potatoes grown under the same conditions, as shown by the number of tubers at various readings.

Variety	Number of Tubers — Specific Gravity												Average Specific Gravity
	1.058	1.062	1.066	1.070	1.074	1.078	1.082	1.086	1.090	1.094	1.098	1.102	
Green Mountain	—	—	—	—	—	—	1	4	7	6	3	1	1.092
Mohawk	—	—	1	2	2	2	1	9	9	17	8	7	1.090
Russet Rural	—	—	—	—	1	2	5	4	6	4	2	2	1.088
Sequoia	—	—	—	2	1	4	1	5	3	—	—	—	1.082
Pioneer Royal	—	—	—	1	6	6	7	8	1	1	—	—	1.081
Houma	—	—	—	2	3	9	6	4	—	—	—	—	1.079
Katahdin	—	2	1	11	10	10	16	14	2	2	—	—	1.079
Chippewa	—	—	2	15	22	12	21	5	1	—	—	—	1.077
Irish Cobbler	—	—	1	2	4	6	1	2	—	—	—	—	1.077
Warba	—	—	—	1	8	4	4	—	—	—	—	—	1.077
Sebago	2	3	1	—	4	3	3	3	1	—	—	—	1.075
Pontiac	6	6	8	20	6	8	2	7	—	—	—	—	1.071
Earlaine No. 2	3	2	3	2	1	1	—	—	—	—	—	—	1.066

"Specific Gravity Determines Potato Mealiness"
Potato Chipper, October, 1948.

specific gravity tubers from a given variety have a greater degree of mealiness when cooked, tend to be dryer, will slough more, and will produce a cooked product of more desirable texture than low specific gravity tubers from the same crop. However, tubers of the same specific gravity from different varieties may cook quite differently.

During processing, french fries or potato chips absorb oil from the fryers. The lower the specific gravity of the raw product, the more oil it absorbs (Lulai and Orr, 1979). The higher rate of oil absorption by lower specific gravity potatoes means that the finished product will have a lower quantity of potato tissue and more oil in each unit of measure of product. With today's concern for eating healthy foods, high specific gravity potatoes which absorb less oil should provide a more desirable finished product (higher yield) from each unit measure of raw potato tubers processed (Greig and Smith, 1960).

How Is Specific Gravity Determined?

There are basically three methods that have been or are being used by the industry for determining the specific gravity of potato tubers. The three methods are (1) hydrometer, (2) weight in air/weight in water, and (3) salt brine solutions.

1. The potato hydrometer was developed for the potato chip industry to provide a rapid method of determining specific gravity of the raw product. The hydrometer consists of a plastic bulb attached to a clear plastic tube with a calibration and measurement sleeve. The plastic tube is closed with a cork. Precisely eight pounds of tuber tissue is placed into a calibrated stainless steel basket. The hydrometer is attached to the basket, the sample and hydrometer are placed in a container of clean water, and a reading is made. The hydrometer is calibrated for use in water which is maintained at 10°C (50°F). If the water is not at this temperature, the specific gravity reading should be adjusted by the amounts listed in Table 23.

2. The weight in air/weight in water method has become the most widely used means of determining specific gravity because of its speed and accuracy. This method requires that a weighing scale be located over a container of water, with room to suspend a nonabsorbent basket containing the sample. An air weight measurement of the tubers is made, followed by a measurement of the weight of the same

TABLE 23. Correction factor for specific gravity of potatoes [corrected to zero at 10°C (50° F) potato temperature and 10° C (50° F) water temperature].

Water Temperature

Tuber Temperature C°	F°	C°3.3 F°38°	4.4 40°	7.2 45°	10.0 50°	12.8 55°	15.6 60°	18.3 65°	21.1 70°	23.9 75°	26.7 80°
3.3	38°	-.0021	-.0020	-.0018	-.0018	-.0020	-.0023	-.0029	-.0038	-.0047	-.0058
4.4	40°	-.0017	-.0016	-.0014	-.0014	-.0016	-.0019	-.0025	-.0034	-.0043	-.0052
7.2	45°	-.0009	-.0008	-.0006	-.0006	-.0008	-.0011	-.0017	-.0026	-.0035	-.0044
10.0	50°	-.0003	-.0002	0000	0000	-.0002	-.0005	-.0011	-.0020	-.0029	-.0038
12.8	55°	+.0001	+.0002	+.0004	+.0004	+.0002	-.0001	-.0007	-.0016	-.0025	-.0034
15.6	60°	+.0004	+.0005	+.0007	+.0007	+.0005	+.0002	-.0004	-.0013	-.0022	-.0031
18.3	65°	+.0005	+.0006	+.0008	+.0008	+.0006	+.0003	-.0003	-.0012	-.0021	-.0030
21.1	70°	+.0007	+.0007	+.0009	+.0009	+.0007	+.0004	-.0002	-.0011	-.0020	-.0029
23.9	75°	+.0007	+.0008	+.0010	+.0010	+.0008	+.0005	-.0001	-.0010	-.0019	-.0028
26.7	80°	+.0008	+.0009	+.0011	+.0011	+.0009	+.0006	0000	-.0009	-.0018	-.0027
29.4	85°	+.0009	+.0010	+.0012	+.0012	+.0010	+.0007	+.0001	-.0008	-.0017	-.0026
32.2	90°	+.0010	+.0011	+.0013	+.0013	+.0011	+.0008	+.0002	-.0007	-.0016	-.0025
35.0	95°	+.0011	+.0012	+.0014	+.0014	+.0012	+.0009	+.0003	-.0006	-.0015	-.0024
37.8	100°	+.0012	+.0013	+.0015	+.0015	+.0013	+.0010	+.0004	-.0005	-.0014	-.0023

United States Standards for Grades of Potatoes for Processing Revised, effective April 14, 1983.

tubers while suspended in the basket submerged completely in the water below the scale. The specific gravity is calculated by the following method:

$$\text{Specific gravity} = \frac{\text{weight in air}}{\text{weight in air} - \text{weight in water}}$$

A correction factor for water or tuber temperatures other than 10°C (50°F) should be made (see Table 23).

3. Specific gravity of potato tubers can be determined by salt brine separation. Salt is usually used to provide brines of differing specific gravity because it is relatively inexpensive, although any number of other substances may be used to create a density solution series. The amount of salt required to make solutions of different densities is given in Table 24. The principle of this method is that an object will float in a solution of density equal to its own. If a potato has a density (specific gravity) of 1.080, then when it is placed in a solution with a density of 1.080, the potato will float. By making up a series of solutions of varying densities, a sample of potatoes can be separated into specific gravity groups.

Tubers are placed in the lowest density solution first. Any potatoes that float in this solution have a density the same as or lower than the density of the solution. All potatoes that sink to the bottom are removed, drained to remove excess water, and then placed in the next higher density solution. A potato that floats in this solution has a specific gravity somewhere between the first and second solution. The procedure is continued for as many solutions as desired. From this determination, not only can an average specific gravity be obtained, but the range and variance of specific gravity is found for the sample. This method is slower and more detailed than the other methods, but provides important information that the others do not normally give. The weight in air/weight in water method can obviously be performed on individual tubers, but generally takes longer than the salt solution method for this type of analysis.

TABLE 24. Amount of salt (NaCl) required per 100 ml of water for brine solutions of densities from 1.0615 to 1.1110.

Salt (NaCl) weight (g/100 ml) (× 1.336 = ounces/gallon)	Density at 6°C (68°F) (specific gravity)	Salt (NaCl) weight (g/100 ml) (× 1.336 = ounces/gallon)	Density at 6°C (68°F) (specific gravity)
8.5	1.0615	11.9	1.0868
8.7	1.0630	12.1	1.0884
8.9	1.0645	12.3	1.0899
9.1	1.0660	12.5	1.0914
9.3	1.0675	12.7	1.0929
9.5	1.0689	12.9	1.0944
9.7	1.0704	13.1	1.0960
9.9	1.0719	13.3	1.0975
10.1	1.0733	13.5	1.0990
10.3	1.0748	13.7	1.1005
10.5	1.0763	13.9	1.1020
10.7	1.0778	14.1	1.1035
10.9	1.0793	14.3	1.1050
11.1	1.0808	14.5	1.1065
11.3	1.0823	14.7	1.1080
11.5	1.0838	14.9	1.1095
11.7	1.0853	15.1	1.1110

Source: *CRC Handbook of Chemistry and Physics*, 47th Edition.

Special Considerations for Specific Gravity Tests

The size of the tubers which make up the sample influences the specific gravity reading. In general, larger tubers have a higher specific gravity. It is important, therefore, that a random sample of all tuber sizes be used.

Potatoes should be clean before measuring the specific gravity. If soil and debris are present in the sample, it will change the specific gravity of the water. Avoid using tubers suspected of having hollow heart; cutting the tubers open will expose any hollow heart cavities.

Be consistent as to whether the tubers in the samples are dry or wet when weighing begins. Wet tubers will have a slightly higher specific gravity (about .001 units) than dry tubers.

The water and tubers used should be at a specified temperature. If this temperature is other than the 10°C (50°F), appropriate correction needs to be made (see Table 23).

Precision in weight determinations are required. A small error in water weights can make a large difference in the specific gravity value obtained.

Agitate the tubers in the basket lightly to remove air bubbles as tubers in the basket are lowered into the water.

Converting Specific Gravity Values to Dry Matter

Since specific gravity is being used as a method to estimate dry matter of the potato, it is appropriate to convert the specific gravity units to dry matter percentage. A number of conversion tables have been published for this purpose. It is important to understand that a potato or lot of potatoes having the same specific gravity may not have the same dry matter content. However, generally speaking, the higher the specific gravity, the higher the dry matter.

A conversion table is generated by making a specific gravity measurement and then oven-drying the same potato at 105°C (220°F) and measuring its dry matter. The two values are then used to identify a specific point on a graph.

After dozens of tubers have been measured in this fashion, a general line can be ascribed to the set of points which is called a regression line. It should be noted that not all data points fall on the line, but the line indicates an approximate dry matter value for a specific gravity value and vice versa.

Regression lines have been determined for different varieties, different years, and different locations. It is not known why variations exist and, for the time being, a representative (acceptable) table must be used.

Measurements of the specific gravity of potato tubers provide an estimate of tuber dry matter content. Tuber dry matter content is a major factor in the processing quality of this crop. The measurement of specific gravity is not absolute and is subject to known, as well as poorly understood, variables.

CONCLUSION

Production of potatoes as a food crop requires knowledge of may facets of agronomy, horticulture, pathology, entomology, engineering, soils, marketing, and business. This crop is one of the more complex in regards to the production system that has evolved around it. Good producers are managers who can assimilate information, utilize resources to their best value, and plan activities to be accomplished in a timely fashion. When good crops have been harvested and sold or placed into storage for future sale, a great deal of satisfaction may be obtained by successful managers.

Currently, new concerns about the effect of crop production on the environment have added a new dimension to managing the potato production system. I believe, however, that if managers utilize the information contained in this text and in the key references cited in it, the negative impacts on the environment will be minimized and the positive benefits to humanity will be significant.

References

Agle, W. M. and G. W. Woodbury. 1968. "Specific gravity–dry matter relationship and reducing sugar changes affected by potato variety, production area and storage." *Amer. Pot. J.* 45:119-131.

Agriculture Statistics. Selected years from 1928-1989. United States Department of Agriculture.

Allen, E. J. 1978. "Plant Density." In *The Potato Crop: The Scientific Basis for Improvement*, edited by P. M. Harris. Chapman and Hall, New York. pp. 279-326.

Arteca, R. N., B. W. Poovaiah, and O. E. Smith. 1979. "Changes in carbon fixation, tuberization, and growth induced CO_2 applications to the root zone of potato plants." *Sci.* 205:1279-1280.

Baker, D. A. and J. Moorby. 1969. "The transport of sugar, water, and ions into developing potato tubers." *Ann. Bot.* 33:729-741.

Bennett-Clark, T. and N. P. Kefford. 1953. "Chromatography of the growth substances in plant extracts." *Nature* 171:645-646.

Binning, L. K., E. A. Kiegel, E. E. Schulte, J. A. Wyman, and W. R. Stevenson. 1980. "Potatoes chemical recommendations." *Univ. Wisc. Comm. Veg. Prod. Bull.* A2352.

Blumenthal-Goldschmidt and L. Rappaport. 1965. "Regulation of bed rest in tubers of potato *Solanum tuberosum* L. VII. Inhibition of sprouting by inhibitor B complex and reversal by gibberellin B complex A3." *Plant Cell Physiol.* 6:601-608.

Boo, L. 1961. "Effect of gibberellic acid on the inhibitor B complex in resting potato." *Physiol. Plant.* 14:676-681.

Borah, M. N. and F. L. Milthorpe. 1962. "Growth of the potato as influenced by temperature." *Indian J. Plant Physiol.* 5:53-72.

Bremner, P. M. and R. W. Radley. 1966. "Studies in potato agronomy. II. The effects of variety, seed size, and spacing on growth, development and yield." *J. Agr. Sci. Camb.* 66:241-252.

Bremner, P. M. and M. A. Taha. 1966. "Studies in potato agronomy. I. The effects of variety, seed size and spacing on growth, development and yield." *J. Agr. Sci. Camb.* 66:241-252.

Brian, P. W., H. G. Hamming, and M. Radley. 1955. "A physiological comparison of gibberellic acid with some auxins." *Physiol. Plant* 8:899-912.

Burt, R. L. 1964. "Carbohydrate utilization as a factor in plant growth." *Aust. J. Biol. Sci.* 17:867-877.

Bushnell, J. 1925. "The relationship of temperature to growth and respiration in the potato plant." *Minn. Agric. Exp. Sta. Tech. Bul.* 34.

Bushnell, J. 1930. "Rate of planting potatoes with some reference to sprouting habit and size of plants." *Ohio Agric. Exp. Sta. Bul.* 462.

Chase, R. W. 1989. *National Potato Council Potato Statistics Yearbook.* Michigan State University.

Clark, Richard S. 1990. *Seed Class Terminology Potato Association of America*; Certification Section.

Claypool, L. L. and O. M. Morris. 1931. "Some responses of potato plants to spacing and thinning." *Proc. Amer. Soc. Hort. Sci.* 253-256.

Correll, D. S. 1962. *The potato and its wild relatives.* Texas Research Foundation, Renner, Texas. 606 pp.

Crafts, A. S. and C. E. Crisp. 1971. *Phloem transport in plants.* Freeman Press, San Francisco.

CRC Handbook of Chemistry and Physics. 1966. Ed. Robert C. Weast, The Chemical Rubber Co. Cleveland OH.

Cutter, E. G. 1978. "Structure and development of the potato plant." In *The Potato Crop: The Scientific Basis for Improvement*, edited by P. M. Harris. Chapman and Hall, New York.

Davis, J. M., W. H. Loescher, M. W. Hammond, and R. E. Thornton. 1986. "Response of potatoes to nitrogen form and to change in nitrogen form at tuber initiation." *J. Amer. Soc. Hort. Sci.* 111:70-72.

Dean, B. B. 1980. "Growth and development of the potato," *Washington State Potato Conference and Trade Fair.* pp. 19-28.

Dean, B. B. 1989a. "Deposition of aliphatic suberin monomers and associated alkanes during aging of *Solanum tuberosum* L. tuber tissue at different temperatures." *Plant Physiol.* 89:1021-1023.

Dean, B. B. 1989b. "Thumbnail checks of potatoes." *Spud Topics.* September.

The Diamond Book: 75th Anniversary of the Northwest Food Processors Association (NWFPA 1989). MEDIAmerica Inc., Portland, OR.

Dow, A. I. 1974. "Fertilizer guide: Irrigated potatoes for central Washington." *Wash. State Agric. Exp. Sta. Bul.* F6-7.

Dunn, L. E. and R. E. Nylund. 1945. "The influence of fertilizers on the specific gravity of potatoes grown in Minnesota." *Amer. Pot. J.* 22:275-288.

Dwelle, R. B. 1985. "Photosynthesis and photosynthate partitioning." In *Potato Physiology*, edited by Paul Li. Academic Press, London. pp. 35-58.

Dwelle, R. B., G. E. Kleinkopf, and J. J. Pavek. 1981. "Stomatal conductance and gross photosynthesis of potato (*Solanum tuberosum* L.) as influenced by irradiance, temperature, and growth stage." *Pot. Res.* 24:49-59.

Edilson, P., R. M. Lister, and W. D. Park. 1983. "Induction and accumulation of major tuber proteins of potato in stems and petioles." *Plant Physiol.* 71:161-168.

El-Antably, H. M. H., P. F. Wareing, and J. Hillman. 1967. "Some physiological responses to D. L. abscision (Dormin)." *Planta* 73:74-90.

Emilsson, B. 1949. "Studies on the rest period and dormant period in the potato tuber." *Acta Agr. Suecava* 3:189-282.

"Fertilizing potatoes." 1977. *Univ. Maine Pot. Inf. Sheet.* No. C-1.

Frier, V. 1977. "The relationship between photosynthesis and tuber growth in *Solanum tuberosum* L." *J. Exp. Bot.* 28:99-107.

Gandar, P. W. and C. B. Tanner. 1976. "Leaf growth, tuber growth, and water potential in potatoes." *Crop Sci.* 16:534-538.

Gardner, E. H., T. L. Jackson, and L. Fitch. 1985. *Oregon State Univ. Fertilizer Guide for Irrigated Potatoes.* FG-57.

Garner, W. W. and H. A. Allard. 1923. "Further studies in photoperiodism, the response of the plant to relative length of day and night." *J. Agr. Res.* 23:871-920.

Gifford, R. M. and J. Moorby. 1967. "The effect of CCC on the initiation of potato tubers." *Eur. Pot. J.* 21:183-193.

Gray, D. 1973. "The growth of individual tubers." *Pot. Res.* 16:80-84.

Gregory, L. E. 1956. "Some factors for tuberization in the potato plant." *Am. J. Bot.* 43:281-288.

Greig, W. Smith, and L. Blakeslee. 1988. "Potatoes: Optimum use and distribution with comparative costs by major regions of the U.S." *EB 1495.* Cooperative Extension, Washington State University.

Greig, W. Smith and O. Smith. 1960. "Factors affecting quality and storage life of prepeeled potatoes and quality of french fries." *Memoir 370.* Cornell University Agric. Exp. Sta., Ithaca, NY.

Gunasena, H. P. M. and P. M. Harris. 1969. "The effect of CCC and nitrogen on the growth and yield of the second early potato variety." *J. Agr. Sci. Camb.* 73:245-259.

Hawkes, J. G. 1978. "History of the Potato." In *The Potato Crop: The Scientific Basis for Improvement,* edited by P. M. Harris. Chapman and Hall, New York, p. 13.

Haynes, R. J. and K. M. Goh. 1978. "Ammonium and nitration of plants." *Biol. Rev.* 53:465-510.

Hayward, H. E. 1938. *The structure of economic plants.* Macmillan Co., New York. pp. 514-549.

Hemberg, T. 1970. "The action of some cytokinins on the rest period and the control of acid growth inhibiting substances in potato." *Physiol. Plant* 23:850-858.

Hemberg, T. 1985. "Potato rest." In *Potato Physiology,* edited by Paul Li. Academic Press Inc., London. pp. 353-388.

Hoff, J. E., S. Lockham, and H. T. Erikson. 1978. "Breeding for high protein and dry matter in the potato at Purdue University." *Purdue Univ. Res. Bul.* 953.

Hooker, W. J. 1981. *Compendium of Potato Diseases.* American Phytopathological Society.

Houghland, G. V. C. 1949. "Nutrient deficiencies in the potato." In *Hunger Signs in Crops,* edited by Howard B. Sprague. David McKay Co. Publ., New York.

Integrated Pest Management for Potatoes in the Western United States. 1986 Western Regional Research Publication 011. University of California, Division of Agriculture and Natural Resources Publication 3316.

Iritani, W. M. 1981. "Growth and preharvest stress and processing quality of potatoes." *Amer. Pot. J.* 58:71-80.

Jones, P. J. and C. G. Painter. 1974. "Tissue analysis. A guide to nitrogen fertilization of Idaho Russet Burbank potatoes." *Univ. of Idaho Current Info. Series.* No. 240.

Kolattukudy, P. E. and B. B. Dean. 1974. "Structure, function and gas chromatographic measurement of suberin from potato tuber slices." *Plant Physiol.* 54:116-121.

Krauss, A. 1978. "Tuberization and abscisic acid content in *Solanum tuberosum* as affected by nitrogen nutrition." *Potato Res.* 21:183-193.

Ku, S., G. Edwards, and C. B. Tanner. 1977. "Effects of light, carbon dioxide and temperature on photosynthesis, oxygen inhibition." *Plant Physiology* 59:868-872.

Kumar, D. and P. F. Wareing. 1972. "Factors controlling stolon development in the potato plant." *New Phytol.* 71:639-648.

Kumar, D. and P. F. Wareing. 1973. "Studies on tuberization on *Solanum andigena*. I. Evidence for the existence and movement of a specific tuberization stimulus." *New Phytol.* 72:283-287.

Kunkel, R. 1966. "Cultural Practices and Their Effect on Potatoes for Processing." Plant Science Symposium. Campbell Institute for Agricultural Research. p. 186.

Kunkel, R. 1969. "Potato crop nutrient removal." *Wash. State Potato Conf. Proc.* pp. 33-42.

Kunkel, R. and W. H. Gardner. 1965. "Potato tuber hydration and its effect on blackspot of Russet Burbank potatoes in the Columbia Basin of Washington." *Am. Potato J.* 42:109-124.

Kunkel, R., N. M. Holstad, and W. H. Gardner. 1977. "Sources of potassium." *Wash. State Pot. Conf. Proc.* pp. 63-69.

Li, P. H. (ed.) 1985. *Potato Physiology.* Academic Press Inc., London.

Lorenz, O. A. 1944. "Studies on potato nutrition. II. Nutrient uptake at various stages of growth by Kern County (Calif.) potatoes." *Proc. Amer. Soc. Hort. Sci.* 44:389-394.

Lorenz, O. A. 1947. "Studies on potato nutrition. III. Chemical composition and uptake of nutrients by Kern County potatoes." *Amer. Potato J.* 24:281-293.

Lulai, E. C. and P. H. Orr. 1979. "Influence of potato specific gravity and yield and oil content of chips." *Amer. Pot. J.* 56:379-390.

Marinus, J. and K. B. A. Bodlaender. 1975. "Response of some potato varieties to temperature." *Potato Res.* 18:189-204.

McCollum, R. E. 1978. "Analysis of potato growth under and allocation of dry matter." *P. Agron. J.* 70:51-57.

Medsger, O. P. 1939. *Edible wild plants.* Macmillan Co., New York. p. 200.

Mingo-Castel, A. M., F. B. Negm, and O. E. Smith. 1974. "The effect of carbon dioxide and ethylene on tuberization of isolated potato stolons cultured *in vitro*." *Plant Physiol.* 53:789-801.

Mingo-Castel, A. M., O. E. Smith, and J. Kumamoto. 1976. "Studies on carbon dioxide promotion and ethylene inhibition of tuberization in potato explants cultured *in vitro.*" *Plant Physiol.* 53:789-801.

Misener, G. C. 1982. "Potato planters–uniformity of spacing." *Trans. Amer. Soc. Agr. Eng.* Vol. 25 No. 6:1504-1505.

Moorby, J. 1968. "The influence of carbohydrate and mineral nutrient supply on the growth of potato tubers." *Ann. Bot.* 32:57-68.

Moorby, J. 1970. "The production, storage and translocation of carbohydrates in developing potato plants." *Ann. Bot.* 34:297-308.

Mulder, E. G. 1949. "Mineral nutrition in relation to the biochemistry and physiology of potatoes. I. Effect of nitrogen, phosphate, potassium, magnesium and copper nutrition on the tyrosine content and tyrosinase activity with particular reference to blackening of the tubers." *Plant and Soil II* 1:59-121.

North American Potato Varieties Handbook. 1990. The Potato Association of America.

Nosberger, J. and E. C. Humphries. 1965. "The influence of removing tubers on dry matter production and net assimilation rate of potato plants." *Ann. Bot.* 29:579-588.

Oparka, K. J., B. Marshall, and D. K. L. Mackerron. 1986. "Carbon partitioning in a potato crop in response to applied nitrogen." In *Phloem Transport*, edited by James Cronshaw, William J. Lucas, and Robert T. Giaquinta. Alan R. Liss, Inc. pp. 577-587.

Painter, C. G. 1979. "Nutrient use by potato vines and tubers." *Univ. Idaho. Current Inf. Ser.* No. 470.

Painter, C. G., J. P. Jones, R. E. McDole, R. D. Johnson, and R. E. Ohms. 1977. "Idaho Fertilizer Guide, Potatoes." *Univ. Idaho Current Inf. Ser.* No. 261.

Palmer, C. E. and O. E. Smith. 1969. "Cytokinins and tuber initiation in the potato *Solanum tuberosum* L." *Nature* 221:279-280.

Pascal, J. A., T. P. Robertson, and D. Longley. 1977. "Yield effects of regularly and irregularly spaced potato tubers." *Exp. Husb.* 32:25-33.

Patterson, D. R. 1975. "Effect on CO_2 enriched internal atmosphere on tuberization and growth of potato." *J. Amer. Soc. Hort. Sci.* 100:431-434.

Pelter, G. Q. 1984. "Columbia Basin Seed Size Survey." *Spud Topics*, 30(21).

Plissy, Edwin S. 1975. *Identification of Potato Aphids.* University of Minnesota and North Dakota State University.

Potato Statistical Yearbook. 1990. The National Potato Council, Englewood, CO.

"Potato Storage and Ventilation." *Washington State University Extension Bulletin,* EM2799.

Rappaport, L. and O. E. Smith. 1962. "Gibberellins in the rest period of the potato tubers." Symposium der Oderhessischen Gesellshaft für Naturund Heilkunde Naturwissenschaftliche Abtielung, zu Giesen vom 1. bis. 3 December 1960, pp. 37-45.

Rappaport, L. and N. Wolf. 1969. "Promotion of dormancy in potato tubers and related structures." *Symp. Soc. Exp. Biol.* 23:219-240.

Rappaport, L., L. F. Lippert, and H. Timm. 1957. "Sprouting, plant growth and tuber production as affected by chemical treatment of white potato seed pieces. I. Breaking the rest period with gibberellic acid." *Amer. Pot. J.* 34:256-260.

Roberts, S. and H. H. Cheng. 1984. "Potato response to rate, time and method of nitrogen application." *Wash. State Pot. Conf. and Trade Fair.* pp. 29-34.

Robins, J. S. and C. E. Domingo. 1956. "Potato yield and tuber shape as affected by severe soil moisture deficits and plant spacing." *Agron. J.* 48:488-492.

Rykbost, K. A., H. L. Carbon, and R. Voss. 1990. *Potato Varieties.* Oregon State University and University of California.

Sharma, K. N. and N. R. Thompson. 1956. "Relationship of starch grain size to specific gravity of potato tubers." *Quarterly Bul. Michigan Agr. Exp. Sta.* 38:559-569.

Shik, C. Y. and L. Rappaport. 1970. "Regulation of bud rest in tubers of potato, *Solanum tuberosum* L. VII. Effect of abscisic and gibberellic acids on nucleic acid synthesis in excised buds." *Plant Physiol.* 45:33-36.

Slater, J. W. 1968. "The effect of night temperature on tuber initiation of the potato." *Eur. Potato J.* 11:14-22.

Smith, O. 1968. *Potatoes: Production, storing, processing.* AVI Publishing. Westport, CT.

Smith, O. E. and L. Rappaport. 1961. "Endogenous gibberellins in resting and sprouting potato tubers." *Adv. Chem. Ser.* 28:42-48.

Smittle, D. A., R. E. Thornton, C. L. Peterson, and B. B. Dean. 1974. "Harvesting potatoes with minimum damage." *Amer. Pot. J.* 51:152-164.

Soueges, R. 1907. Development et structure la tegument seminal chez les solanaces. *Ann. Sci. Nat. Bot. Ser.* IX (6):1-124.

"Specific Gravity Determines Potato Mealiness" 1948. *Potato Chipper.*

Stall, W. M. and M. Sherman. "Potato production in Florida." *Florida Coop. Ext. Circ.* 118.

Struik, P. C., B. J. Schnieders, L. H. Kerckhoffs, and G. W. J. Visscher. 1988. "A device for measuring the growth of individual potato tubers non-destructively and precisely." *Pot. Res.* 31:137-143.

Svenson, B. 1977. "Changes in seed tubers after planting." *Pot. Res.* 20:215-218.

Thomas, W. and W. B. Mack. 1938. "Foliar diagnosis in relation to development and fertilizer treatment of the potato." *J. Agric. Res., Comb.* 57:397-414.

Thornton, R. E., D. A. Smittle, and C. L. Peterson. 1974. "Reducing potato damage during harvest." *Wash. State Univ. Ext. Bul.* 646.

Tizio, R. 1969. "Action du CCC (chlorure de [2-chloroethyl]-trimethlammonium) suf la tuberisation de la de terre." *Eur. Pot. J.* 12:307.

United States Standards for Fresh Potatoes. 1972. U. S. Dept. of Agric.

United States Standards for Grades of Potatoes for Processing. 1983. United States Department of Agriculture.

Vitosh, M. L. 1990. "Potato Fertilizer Recommendations." *Mich. State Univ. Ext. Bul.* E2220.

Wagner, D. F., W. C. Dahuke, D. C. Nelson, and E. H. Vasey. 1977. "Fertilizing potatoes." *North Dakota State Univ. Cic.* S-F6.

Weaver, M. L., H. Timm, M. Nonaka, R. N. Sayre, R. M. Reeve, R. M. McCready, and L. C. Whitehand. 1978. "Potato Composition I: Tissue selection and its effects on solids content and amylose/amylopectin ratios." *Amer. Pot. J.* 55:73-82.

Winkler, E. 1971. "Kartoffelbau in Tirol. II. Photosynthesis-vermogen und respiration von verschiedenen kartoffelsorten." *Pot. Res.* 13:1-18.

Yamaguchi, M., H. Timm, and A. K. Spurr. 1964. "Effects of soil temperature on growth and nutrition of potato plants and tuberization, composition and periderm structure of tubers." *Proc. Amer. Soc. Hort. Sci.* 84:412-423.

ADDITIONAL READING

Burton, W. G. 1989. *The Potato*, 3rd Edition. Halsted Press. New York.

Cargill, B. F. 1986. *Engineering for potatoes.* Michigan State University and American Society of Agricultural Engineers.

Flint, M. L., L. L. Strand, P. A. Rude, and J. K. Clark. *Integrated Pest Management for Potatoes in the Western United States.* 1986. University of California, Division of Agriculture and Natural Resources Publication 3316. Western Regional Research Publication 011.

Harris, P. M. (ed) 1978. *The Potato Crop: The Scientific Basis for Improvement.* Chapman & Hall. New York.

Hollingsworth, C. S., D. N. Ferro, and W. M. Coli. 1986. *Potato Production in the Northeast: A Guide to Integrated Pest Management.* Univ. of Mass. Publication.

Hooker, W. J. 1981. *Compendium of Potato Diseases.* American Phytopathological Society. St. Paul, MN.

Smith, O. 1968. *Potatoes: Production, Storing, Processing.* AVI Publishing Co. Westport, CT.

Talburt, W. and O. Smith. 1967. *Potato Processing.* AVI Publishing Co. Westport, CT.

APPENDIX

Form 8203

Universal Seed Potato Contract.

1. DATE OF ISSUE _____

2. ORIGINAL ISSUER: _____ Seller
   (Check one) _____ Buyer

3 CONTRACT NO. 3824

This is a legal CONTRACT FOR SALE for either a present or future sale in consideration of the mutual promises hereinafter. between the SELLER.

Seller's Business Name	Address	City	State	Zip	Telephone Number

hereinafter referred to as SELLER, who has the legal authority to sell or offer to sell the hereinafter described seed potatoes (either crops, farm products or goods) and agrees to sell according to all the terms and conditions of this agreement, WHICH SHALL BE NULL AND VOID UNLESS SIGNED BY BOTH PARTIES HERETO AND RETURNED TO THE ORIGINAL ISSUER WITHIN FIFTEEN (15) DAYS AFTER THE HEREABOVE CONTRACT DATE OF ISSUE (ITEM NO. 1), **PROVIDED HOWEVER,** NOTWITHSTANDING ITS ENFORCEMENT AND FULL EFFECT WITH RESPECT TO GOODS FOR WHICH ANY PAYMENT HAS BEEN MADE AND ACCEPTED, OR WHICH GOODS HAVE BEEN RECEIVED AND ACCEPTED, and the BUYER,

Buyer's Business Name	Address	City	State	Zip	Telephone Number

hereinafter referred to as BUYER, who hereby agrees to accept and pay for in-full for the hereinafter described seed potatoes (either crops, farm products or goods) according to all the terms and conditions of this agreement, including **REVERSE SIDE HERETO,** the FULL PAYMENT PROMPTLY CLAUSE, the **LATE PAYMENT CLAUSE,** the **LIMITATION AND EXCLUSION OF CERTAIN WARRANTIES,** the **LIMITATION OF CONSEQUENTIAL DAMAGES AND REMEDIES,** the MISCELLANEOUS CONDITIONS OF SALE, and all of the other provisions hereto as follows:

4. SEED STATE OF ORIGIN	5. POTATO VARIETY or NO.	6. FIELD/LOT IDENTIFICATION	7. SEED CLASS/GENERATION
8. TAG COLOR	9. STATE SEED GRADE	10. MINIMUM TUBER SIZE	11. MAXIMUM TUBER SIZE
12. TRADE TERMS OF SALE: (See P.A.C.A. Regulation Sec. 46.43.) SOLD AS:			13. ____ BAG SIZE ____ lbs. ____ BULK SHIPMENT
14. TYPE OF CARRIER:	15. QUANTITY of CWT CWT	16. F.O.B. UNIT PRICE per CWT $ /CWT	17. TOTAL F.O.B. PURCHASE PRICE $
18. TERMS OF DISCOUNT:		18(a). LESS DISCOUNT per CWT $ /CWT	18(b). LESS TOTAL DISCOUNT $

19. FREIGHT: Arrangements and Responsibilities:	19(a). FREIGHT COST per CWT	19(b). TOTAL FREIGHT COST
	$ /CWT	$
	20(a). TOTAL COST per CWT	20(b). TOTAL AMOUNT
	$ /CWT	$
21. The agreed DOWN PAYMENT SCHEDULE—	22. 1st Down Payment per CWT $ /CWT	22(a). DATE DUE 1st Down	22(b). 1st Down Payment Amount $
	23. 2nd Down Payment per CWT $ /CWT	23(a). DATE DUE 2nd Down	23(b). 2nd Down Payment Amount $
			24. BALANCE DUE $

24. **BALANCE DUE** — FULL PAYMENT PROMPTLY —
Full payment is due within **TEN (10) DAYS** after the day on which the said seed potatoes are accepted at delivery by the receiver with respect to each tender or within **TEN (10) DAYS** after the Date of Invoice, whichever is the later;

25. The agreed SHIPPING-POINT is — _____ ;

26. The agreed DELIVERY-POINT is — _____ ;

27. The agreed SHIPPING-DATE(S), as planned, is — _____ , subject to available transportation and related weather, and the total and final shipment shall be completed no later than _____ ;

NOTICE TO BUYER & SELLER: CONTRACT TERMS AND CONDITIONS CONTINUE ON REVERSE SIDE HERETO AND ARE PART OF THIS AGREEMENT WITH FULL-FORCE AND EFFECT.

SELLER hereby agrees that he has READ BOTH SIDES OF AGREEMENT and agrees to and accept the terms and conditions hereto, including the EXPRESSED REPRESENTATIONS, RESPONSIBILITIES TO PERFORM AND REMEDIES as set forth.

Seller's Signature _____

Business Name _____

28. DATE OF ACCEPTANCE _____

BUYER hereby acknowledges that he has READ BOTH SIDES OF THIS AGREEMENT and agrees to and accepts the terms and conditions hereto, including the LIMITATION OF WARRANTIES, CONSEQUENTIAL DAMAGES AND REMEDIES as set forth.

Buyer's Signature _____

Business Name _____

28. DATE OF ACCEPTANCE _____

Form 8203

Universal Seed Potato Contract (continued)

REVERSE SIDE OF THE UNIVERSAL SEED POTATO CONTRACT FOR SALE

TERMS AND CONDITIONS CONTINUED FROM FRONT SIDE — WITH FULL-FORCE AND EFFECT —

29. The agreed **LATE PAYMENT CHARGES** —

Notwithstanding the laws of any State or Federal law with respect to late payment charges, all accounts not paid-in-full as defined by the hereabove Item No. 24, will be subject to a **LATE PAYMENT CHARGE** equal to the maximum amount or rate permitted by law. If not otherwise prohibited, a **LATE PAYMENT CHARGE** will be imposed on the unpaid balance of all past due accounts, less the amount of payments received and any credits, at an **ANNUAL PERCENTAGE RATE OF TWENTY-ONE PERCENT (21%)**, this is a periodic rate of 0.0575 percent per calendar day, calculated ten (10) days after the day on which the said seed potatoes are accepted at delivery by the receiver with respect to each tender or within ten (10) days after the Date of Invoice, whichever is the later, and until full-payment and all billed **LATE PAYMENT CHARGES** have been received by the seller. If full-payment is received within the 10 days after the said seed potatoes were accepted per tender or as otherwise provided, no **LATE PAYMENT CHARGES** will be imposed. Payments are applied to billed **LATE PAYMENT CHARGES**, next to old purchases, and then to newly unbilled purchases. If twenty-one percent (21%) is not allowed by the respective State law, then the maximum permitted amount or rate will be imposed accordingly.

30. The agreed **EXCLUSION AND LIMITATION OF CERTAIN WARRANTIES** —

Due to the fact that seed potatoes are: perishable vegetative tuber-seeds; unstable under certain conditions; easily contaminated or damaged through handling, shipment, storage, cutting, treating or planting; devitalized or weakened by mishandling or planting during unfavorable or moisture conditions. And because the handling, use, sanitation, cropping, germination, quality after shipping, and physical possession of the seeds are far beyond the control of the producer, SELLER, shipper or regulatory inspectors, including the Federal-State Inspection Service, State Seed Certification Agency, State Department of Agriculture, the following **EXCLUDED AND LIMITED WARRANTIES ARE OFFERED FOR THE SEED POTATOES SOLD BY THIS AGREEMENT:**

   a). The SELLER and the producer represent that the seed potatoes sold and to the label (seed tag) description as required by the Seed State of Origin and/or Federal-State Inspection Laws, and will conform to the requirements specified in the above Items Number 4 through 13 of this agreement; and

   b). **THERE ARE NO WARRANTIES WHICH EXTEND BEYOND THE DESCRIPTION OF THE FACE HEREOF. THE SELLER AND THE PRODUCER MAKE NO OTHER WARRANTIES, EXPRESSED OR IMPLIED, OF MERCHANTABILITY, FITNESS FOR A PARTICULAR PURPOSE, FREEDOM FROM ANY LATENT POTATO DISEASE, VIRUS OR DISORDER OF ANY NATURE, OR OTHERWISE, AND IN ANY EVENT LIABILITY FOR BREACH OF ANY WARRANTY OR CONTRACT WITH RESPECT TO SUCH SEEDS IS LIMITED TO THE ACTUAL PURCHASE PRICE, AS PER TRADE TERMS OF SALE, ITEM NO. 12 HERE-ABOVE.**

31. The agreed **LIMITATION OF CONSEQUENTIAL DAMAGES AND REMEDIES** —

ANY DAMAGES ARISING OUT OF THIS CONTRACT SHALL BE LIMITED IN ALL EVENTS TO THE RETURN OF THE ACTUAL PURCHASE PRICE PAID AS PER TRADE TERMS OF SALE FOR SUCH SEEDS ON THAT PORTION OF THE SEED POTATOES ON WHICH A COMPLAINT MAY ARISE. THE SELLER OR PRODUCER SHALL NOT BE LIABLE FOR PROSPECTIVE PROFITS OR SPECIAL, INDIRECT, OR CONSEQUENTIAL DAMAGES. THE RETURN OF THE ACTUAL PURCHASE PRICE PAID AS PER TRADE TERMS OF SALE FOR SUCH SEEDS IS THE EXCLUSIVE AND SOLE REMEDY AVAILABLE TO THE BUYER OR USER OF THESE SEED POTATOES.

32. The agreed **MISCELLANEOUS CONDITIONS OF SALE —**

a). **CURRENCY.** All prices, amounts and payments herein refer to and are payable in UNITED STATE OF AMERICA DOLLARS;

b). **TRADE TERMS AND DEFINITIONS.** The P.A.C.A. "definitions" (Sec. 46.2) and "trade terms and definitions" (Sec. 46.43) of the official Regulations Under the Perishable Agricultural Commodities Act of 1930 as amended (issued under Sec. 15, 46 Stat. 537; 7 U.S.C. 499o) in effect at the time of the Date of issuance of this agreement, shall be part of this CONTRACT FOR SALE and shall be binding upon the parties and have full-force and effect. The SELLER and the BUYER are both chargeable with having knowledge and skill peculiar to the said sections of the P.A.C.A. Regulations;

c). **FUTURE SALE BASED UPON WINTER TEST RESULTS.** When the future sale of the said seed potatoes are subject to certain post-harvest winter test results or inspection tolerances, and should the said seed potatoes fail such test or tolerances, then the BUYER shall have the right to terminate and end this agreement without breach or recourse against either party, and furthermore, the BUYER shall have the right to receive full refund promptly, without interest, upon request for any and all money paid for the said seed potatoes;

d). **LIMITATION OF PERFORMANCE—REFUND.** The SELLER or producer shall not be liable, or breach, except to fully refund promptly, without interest, any and all money received as partial or full-payment for the said seed potatoes, for failure to perform or deliver tender pursuant to this agreement when such failure is caused by: (1) crop rejection or disqualification by an official State Seed Certification Agency whatever the cause; (2) shipping or transportation related stoppages, damages, strikes, accidents or mishaps, fires, thefts, or failure to perform as per contract-of-haul; or (3) Acts of God or natural disasters, storms, floods, frost or freeze, hail, disease, virus, pathogens, fires or other circumstances beyond the SELLERS or producers control relating to the said seed potatoes, either as crops, farm products in storage, or goods;

e). **OFF-SIZE TOLERANCES.** Notwithstanding the Seed State of Origin regulation regarding off-size tolerances for the respective grade (Item 9 above), and if not otherwise agreed, (1) FOR UNDERSIZE: 3% for potatoes in any load lot which fail to meet the required or specified minimum size except that 5% shall be allowed when the minimum size specified is 2¼ inches or more in diameter, or 5 ounces in weight; and (2) FOR OVERSIZE: 10% for potatoes in any load lot which fail to meet the required or specified maximum size;

f). **WEIGHT TOLERANCES.** Shipments may vary as to actual weight per load, or contract, because of carrier weight limitations, in such case, the final balance due (and freight) shall be determined by multiplying the "actual shipping-point CWT weight" times the "price per CWT"; and

g). **DELIVERY IN LOAD LOTS.** Unless otherwise agreed, it is agreed and understood that shipment and delivery of the said seed potatoes shall be tendered in one or more truck/load or railcar/load lots throughout the agreed SHIPPING DATE(S) and until the total quantity is tendered. Each shipment/delivery shall be the equivalent of a separate contract. Payment may be demanded for one or more load lots upon acceptance as per this agreement or as a single payment upon final invoice as per this agreement.

33. The agreed **STATUTE OF LIMITATION—TERRITORIAL APPLICATION —**

a). It is agreed that any action for breach of this CONTRACT FOR SALE or of any warranty, express or implied, must be commenced within one (1) year after the cause of action has occurred; and

b). It is agreed that the laws of the "Seed State of Origin" as specified hereabove in Item No. 4, and none other, govern this agreement, sales transaction and seed product. If any legal action is brought, the agreed place of venue is to be the "Seed State of Origin."

34. **SUCCESSORS AND ASSIGNS AGREEMENT —**

This agreement shall be binding and have full-force and effect on the heirs, executors, administrators, successors and assigns of the parties hereto.

35. **SEVERABILITY AGREEMENT —**

If a part of this contract is found invalid by a court of law, all valid parts that are severable from the invalid parts remain in effect. If a part of this contract is invalid in one or more of its applications, the part remains in effect in all valid applications that are severable from the invalid applications.

Seed Cutter Evaluation Worksheet

Collect 4.5 kg (10 pounds) of seed from the output conveyor of the seed cutter before any sizing or hand sorting is done. Weigh each seed piece and record the number in each size category below.

Number of seed pieces in each weight range

<28 grams < 1 oz.	28.1-35 grams 1.10-1.25 oz.	35.1-42 grams 1.26-1.50 oz.	42.1-49 grams 1.51-1.75 oz.	49.1-56 grams 1.76-2.0 oz.	56.1-63 grams 2.01-2.25 oz.	>63 grams >2.25 oz.

Calculate the percentage by number of seed pieces in each category. Depending on what the desired size is, adjust the cutter to maximize the number of seed in the desired sizes. The desired sizes are usually 42-56 grams or 1.5-2.0 ounces depending on the cost of seed and the length of the growing season.

SEED PLACEMENT EVALUATION WORKSHEET

Planter performance should be evaluated in the following manner:

1) Uncover the distance of row for each row across the planter width equal to the distance in which 10 seed pieces should be placed. For example: 9" spacing requires 10 seed pieces in 90" or 7.5 feet or for 23 cm spacing there will be 10 seed pieces in 2.3 meters.

2) Record the location of each seed piece found in the appropriate distance of row on the seed placement form.

3) The seed piece counts as a double if it is less than 50% of the distance away from the previously recorded seed piece in regards to the desired distance. A skip is recorded if no seed piece is found before reaching a distance of 50% greater than that desired. The following is an example:

The desired spacing = 9" or 23 cm
Desired seed x x x x x x x x x x x x
Actual seed   0   00 0 0 0   0 0   0 0
This example contains 1 double and 2 skips.

4) A percent accuracy value for a 6-row planter can be calculated as follows:

$$\text{accuracy} = 100 \times \left[\left\{ 60 - \frac{(Q+S+D)}{2} \right\} + 60 \right]$$

   where

Q = The absolute number of pieces (10 – # of pieces) in each row (add the number for each row together)
S = # of skips
D = # of doubles

This value can be used to approximate the accuracy of seed placement in the field.

Seed Placement Form

Place an x where a seed is found in each row across the width of the planter.

0	0	0	0	0	0	0	0	0	0
0	0	0	0	0	0	0	0	0	0
0	0	0	0	0	0	0	0	0	0
0	0	0	0	0	0	0	0	0	0
0	0	0	0	0	0	0	0	0	0
0	0	0	0	0	0	0	0	0	0
0	0	0	0	0	0	0	0	0	0
0	0	0	0	0	0	0	0	0	0
0	0	0	0	0	0	0	0	0	0
0	0	0	0	0	0	0	0	0	0

Field Monitoring Form for Insects, Diseases and Disorders

FIELD: _____ VARIETY _____ DATE: _____

Area of Sampling   COMMENTS: _____

N

_____

_____

_____

_____

_____

S

Location in the Field

	Area 1	Area 2	Area 3	Area 4	Area 5
CHECK 5 PLANTS & RECORD AVG. NO. INSECTS PER PLANT:					
Aphids					
Loopers					
Potato Beetles					
Wireworms					
Nematodes					
CHECK 5 PLANTS & RECORD # OF PLANTS WITH DISEASE:					
Rhizoctonia					
Black leg					
Mosaic					
Leafroll					
Early Blight					
Late Blight					
Scab					
Other _____					
DISORDERS FOR 2 HILLS DUG:					
Hollow Heart					
Brown Center					
Heat Necrosis (IBS)					
Vascular Discoloration					
Lenticels					

SEED CERTIFICATION AGENCIES

Alaska Plant Materials Center
  HC02 Box 7440
  Palmer, AK 99645
  (907) 745-4469

British Columbia Interior Seed Potato Producers
  1130 Victoria Street
  Kamloops, B. C.
  Canada V2C 2C5
  (604) 375-2346

Colorado Certified Potato Growers Association
  Potato Certification Service
  San Luis Valley Research Center
  0249 E. Rd. 9N
  Center, CO 81125
  (719) 754-3496

Idaho Crop Improvement Association
  P.O. Box 51139
  Idaho Falls, ID 83405
  (208) 522-9198

Maine Certified Seed
  Division of Plant Industry
  744 Main Street Suite 9
  Presque Isle, ME 04769
  (207) 764-2036

Michigan Crop Inprovement Association
  P.O. Box 21008
  Lansing, MI 48909
  (517) 355-7438

Minnesota Seed Potato Certification
  12 Hill Hall
  University of Minnesota
  Crookston, MN 56717
  (218) 237-6976

Montana Certified Seed Potatoes
   Potato Plant Growth Center
   Montana State University
   Bozeman, MT 49717
   (406) 994-3150

Nebraska Potato Certification Association
   Box 339
   Alliance, NE 69301
   (308) 762-1674

Nevada Department of Agriculture
   Division of Plant Industry
   P.O. Box 11100
   Reno, NV 89501
   (702) 789-0180

New Brunswick
   Canadian Seed Potato Export Agency
   Suite 102, 191 Prospect Street
   Fredericton, N.B.
   Canada E3B 277
   (506) 453-0788

New York Seed Improvement Cooperative, Inc.
   P.O. Box 218
   Ithaca, New York 14851
   (607) 257-2233

North Dakota State Seed Department
   University Station
   Box 5257
   Fargo, ND 58105
   (701) 237-7927

Oregon Seed Certification
   31 Crop Science Building
   Oregon State University
   Corvallis, OR 97331-3003
   (503) 737-4513

Pennsylvania Department of Agriculture
    Bureau of Plant Industry
    Agric. Bldg. Room 105
    2301 N. Cameron Street
    Harrisburg, PA 17120
    (717) 787-4843

Prince Edward Island Department of Agriculture
    Plant Industry Services Branch
    Potato Extension Specialist
    P.O. Box 1600
    Charlottetown, Prince Edward Island
    Canada C1A 7N3
    (902) 892-5465

South Dakota Potato Growers Association
    P.O. Box 127
    Clark, SD 57225
    (605) 532-3311

Utah Crop Improvement Association
    Utah State University
    Logan, UT 84322-4855
    (801) 750-2082

Vermont Department of Agriculture
    Division of Plant Pest Control
    Montpelier, VT 05602
    (802) 828-2431

Washington State Seed Potato Commission
    P.O. Box 286
    Lynden, WA 98264
    (206) 354-4670

Wisconsin Seed Potato Certification Agency
    2960 Neva Road
    P.O. Box 328
    Antigo, WI 54409

Wyoming Agricultural Extension Service
  University Station
  P.O. Box 3354
  Laramie, WY 82071
  (307) 766-2248

Index